T0361834

First Order Phase Transitions of Magnetic Materials

Broad and Interrupted Transitions

First Order Phase Transitions of Magnetic Materials

Broad and Interrupted Transitions

Praveen Chaddah

CRC Press
Taylor & Francis Group
Boca Raton London New York

CRC Press is an imprint of the
Taylor & Francis Group, an **informa** business

CRC Press
Taylor & Francis Group
6000 Broken Sound Parkway NW, Suite 300
Boca Raton, FL 33487-2742

© 2018 by Taylor & Francis Group, LLC
CRC Press is an imprint of Taylor & Francis Group, an Informa business

International Standard Book Number-13: 978-1-4987-8625-6 (Hardback)
International Standard Book Number-13: 978-1-315-15588-3 (eBook)

Library of Congress Cataloging-in-Publication Data

Names: Chaddah, Praveen, 1951- author.
Title: First order phase transitions of magnetic materials : broad and
interrupted transitions / Praveen Chaddah.
Description: Boca Raton, FL : CRC Press, Taylor & Francis Group, [2018] |
Includes bibliographical references and index.
Identifiers: LCCN 2017034226| ISBN 9781498786256 (hardback) | ISBN 1498786251
(hardback) | ISBN 9781315155883 (ebook) | ISBN 1315155885 (ebook)
Subjects: LCSH: Phase transformations (Statistical physics) | Magnetic
materials.
Classification: LCC QC175.16.P5 C45 2018 | DDC 530.4/74--dc23
LC record available at https://lccn.loc.gov/2017034226

Visit the Taylor & Francis Web site at
http://www.taylorandfrancis.com

and the CRC Press Web site at
http://www.crcpress.com

Contents

Preface

First order phase transitions abound in nature. The conversion of liquid water into vapor, and its subsequent condensation into liquid, is a sequence of two first order transitions that sustain life on earth. The melting of glacial ice into water is another first order transition that helps sustain life, but the excess of which may threaten life on earth. These first order phase transitions are commonly observed to occur under heating or cooling and are accompanied by sharp discontinuities in physical properties. First order phase transitions are characterized by the temperature change being arrested at the transition temperature T_C, even as heat is being added or removed, until one phase converts fully to the other. The amount of heat added (or removed) without change in temperature at T_C is called latent heat. First order phase transitions are also characterized by a jump in volume ΔV or a jump in magnetization ΔM, depending on whether T_C is affected by a change in pressure P or magnetic field H. The slope of the phase transition line $T_C(P)$ [or $T_C(H)$] is related to the latent heat L and ΔV [or ΔM] through the Clausius–Clapeyron relation and has been an essential experimental test that must be satisfied for a transition to be accepted as a first order phase transition.

This conventional wisdom of what characterizes a first order transition has changed in recent years, mainly due to studies on first order phase transitions of magnetic materials. Over the past two decades, it has been established that many magnetic materials show first order transitions that are broad (they occur over a broad region of temperature and of magnetic field) without sharp discontinuities or latent heat. These broad transitions show hysteresis and are even interrupted midway through the transition. We explain the puzzling observation of phase coexistence of equilibrium low entropy and kinetically arrested high entropy states that persist to the lowest temperatures.

The easy availability of superconducting magnets has led to widespread temperature-dependent measurements of material properties at high magnetic fields. Commercially available equipment has also made isothermal measurements under varying magnetic fields very common, without an emphatic realization that similar isothermal measurements under varying pressures were not possible at varying temperatures. Since T_C of most first order phase transitions is influenced by P, and since varying P at an arbitrary T is experimentally difficult, the phase transition line was traditionally determined by varying T at various fixed values of P. This has changed drastically with studies on phase transitions of superconductors and of magnetic materials because H can now be easily varied isothermally at any arbitrary T. Such new studies have naturally focused on materials whose properties vary with magnetic field, with recent emphasis being on phase transitions in magnetic materials.

First order magnetic phase transitions gained prominence about two decades ago, when transitions were reported between the competing ferromagnetic and antiferromagnetic ordered phases in some half-doped manganites. These transitions also involved a sharp change in resistance, including from an insulating to a metallic state, raising possibilities of applications. Isothermal variations of the second control variable H are very common in studies on such magnetic materials. These have provided general results on first order phase transitions that were not available from earlier studies (using pressure) on nonmagnetic materials and are the raison d'être for this book.

First order phase transitions can exhibit hysteresis upon cooling and warming, with the transition occurring at $T < T_C$ on cooling and at $T > T_C$ on warming. This has been common wisdom but was neither thoroughly investigated nor asserted with any rigor in the traditional Ehrenfest classification of phase transitions. As we shall discuss in detail, the simplistic but exactly solvable van der Waals model for the liquid–gas phase transition gives exact solutions for such hysteresis. The limits on the extent of supercooling and superheating that cause this hysteresis are obtained analytically in this model, even though the phase transition temperature can only be obtained numerically! We shall discuss this prediction of the existence of the corresponding $T^*(P)$ and $T^{**}(P)$ lines that define the limits to which metastable states can exist in this model as a prelude to the importance metastable states hold in experimental studies on first order transitions in magnetic materials.

The traditional Ehrenfest classification of phase transitions was based solely on what integer derivative of the free energy showed a discontinuity at T_C and became inapplicable for many phase transitions where the specific heat was continuous but approached infinity at T_C with opposing slopes; the function was continuous but its derivative could not be defined. This inability to classify many observed phase transitions led to the so-called modern classification wherein phase transitions were classified either as first order or as continuous. This development also led to the conclusion that only first order phase transitions can show hysteresis in measurement of T_C, providing a defining observation for first order transitions. We shall discuss this in some detail, considering how metastable states associated with the hysteretic behavior can be converted to the stable state by varying the second control variable, viz., pressure or magnetic field. This enables ensuring that the observed hysteresis is not due to an experimental artifact like thermal lag nor due to arrested kinetics as in the pinning of magnetic domains. The stringent requirements for establishing compliance with the Clausius–Clapeyron equation pose experimental difficulties when the latent heat L is small, and it becomes difficult even to distinguish it from a peak in specific heat [1].

The genesis of this book can be traced to developments a little over two decades ago. There were three interesting phenomena studied in the mid-1990s that were being attributed to underlying phase transitions, which were also caused by varying magnetic field and were possibly first order. Each of these required measurements at low temperatures and in high magnetic fields, and the latent heats were difficult to measure. The underlying phase transitions, and the materials being studied, were (1) the melting of the vortex lattice in the high-T_C superconductors, (2) the onset of the peak-effect in the critical current density J_C of superconductors, and (3) the melting of "charge-order" across the antiferromagnetic insulator (AFI) to ferromagnetic metal (FMM) transition in the half-doped manganites. The studies on the first phenomenon were motivated by experimental reports showing hysteresis at the jump in resistance across the transition, but resistance is not a derivative of free energy. It is not an equilibrium physical property, and a discontinuity in it lacks significance in the Ehrenfest classification. This was firmly established as a first order transition only in 1996 following measurement of latent heat and by showing that the Clausius–Clapeyron relation was satisfied all along the phase transition line in (H, T) space of control variables [2].

The studies on the second phenomenon were motivated by hysteresis in J_C, which is not an equilibrium property but is related to hysteresis in the equilibrium property of magnetization. In this case, there were attempts to extract the jump in equilibrium magnetization and then use the Clausius–Clapeyron relation to estimate the latent heat [3,4]. The conclusion that it was a first order transition was supported by showing that the extent of hysteresis was dependent, in a certain predicted manner, on the path followed in the (H, T) space [5]. The latent heat was not directly measured, and the Clausius–Clapeyron relation was not established for this phenomenon.

The report on the third phenomenon showed sharp jumps in magnetization and resistance and relied on hysteresis in the nonequilibrium property of resistance to assert that this was a first order phase transition [6]; there was no reference to the Clausius–Clapeyron relation in that paper! This report actually appeared before the detailed (and painstaking) report on vortex-lattice melting [2] but presented the shift in emphasis from latent heat to hysteresis envisaged by White and Geballe [1] about two decades earlier. We emphasize that hysteresis could even be in a nonequilibrium physical property.

This shift in emphasis from discontinuous jumps in equilibrium physical properties to hysteresis in even nonequilibrium properties becomes a necessity when we discuss magnetic materials where there is a disorder broadening of the transition. This book provides a detailed basis for this subtle but major shift in the experimental characteristics for identifying first order transitions. The charge-melting transition also showed the coexistence of AFI and FMM phases, with the intriguing feature that this phase coexistence persisted to the lowest temperatures. We also discuss how this led to a major development in our understanding of first order transitions in actual magnetic materials.

Magnetic shape-memory alloys, which have immense potential for applications, usually also show associated magnetic-field-induced first order phase transitions, so do materials showing large magnetoresistance, or large magnetocaloric effect, or large magnetoelectric effect. Another conceptual advance in our understanding of first order phase transitions has been triggered by most of these functional magnetic materials being multicomponent systems whose properties have been made more interesting by substitutions. Such substitutions are an intrinsic source of disorder. Over two decades after the theoretical work of Imry and Wortis [7], it has been experimentally established that in the presence of weak disorder the first order transition occurs over a range of temperatures, with the two competing phases coexisting in different regions of the sample over this broader temperature range. This was also established experimentally when the control variable used was magnetic field instead of temperature. Confirmation that these transitions were first order was provided by hysteresis across the broad transition. Finally, the behavior of the metastable supercooled and superheated states followed theoretical predictions. The identification of such metastable states becomes necessary when disorder broadens a first order phase transition, because the discontinuities (in entropy and magnetization) are now smudged out and cannot be measured. The paradigm shift from Clausius–Clapeyron equation to hysteresis (with metastable to stable transformations under appropriate variations of the second control variable) as a confirmation for a first order phase transition has now become a necessity. We emphasize that such variations are experimentally not possible with P as the second control variable but are so easy with H. Magnetic first order transitions and studies on magnetic materials have thus been essential to the development of these new insights.

Broad first order transitions, occurring under LTHM (low temperature, high magnetic field) conditions, provide a proximity to conditions under which glass-like arrest of kinetics is possible. The proximity of these two can be varied using the second control variable, as has been clearly established in many first order magnetic transitions. This results in many new physical observations, which form a new and unique feature of this book. The first was the observation of coexisting phases down to the lowest temperature, as first noted in the melting of charge order, raising the possibility of an inhomogeneous ground state [8]. Since first order transitions occur over a range of temperatures, there is an interesting possibility that a glass-like arrest occurs within this range and then the transition is arrested when it is partial and incomplete. This possible explanation for phase coexistence persisting to the lowest temperature was proposed by us [9], suggesting a coexistence of a glass-like higher entropy kinetically arrested phase with a crystal-like equilibrium phase. Using different magnetic fields to control the separation of temperatures corresponding to the phase transition and to the glass-like arrest of kinetics allowed tuning these coexisting fractions [10]. The multiplicity of states that could be experimentally obtained at the same temperature and magnetic field clearly brought out that these inhomogeneous states were not equilibrium states [11]. Multiple values of magnetization and resistivity were obtained at the same value of temperature and magnetic field over a range of low temperatures. These properties could be varied over a range of values in a controllable manner, as was predicted by our concept of kinetic arrest. This data provided a challenge to theorists, and it was probably the visually drastic data that helped establish the validity of our new concepts.

The subsequent development of the new protocol of cooling and heating in unequal fields (CHUF) [12], which clearly shows the devitrification of the arrested state and gives unambiguous and rather visual evidence of the coexisting phases being a glass-like arrested state, shall be discussed in detail. CHUF thus allows the measurement of the temperature at which the kinetics is arrested (or de-arrested), and one can study the variation of T_g with H in an exhaustive way. This contrasts with the limited results so far on the pressure dependence of T_g. In addition, for broad transitions, devitrification can be caused to occur at different temperatures by cooling in different fields but warming in the same field. This can further our understanding of nucleation and growth. There is a wide scope for application of the concepts and results presented in this book.

This book discusses new concepts introduced during the last two decades and experimental results that either generated those new concepts or established conclusions that followed from them. Schematic figures that describe concepts are presented in the simplest form that captures the essence of the concept while recognizing that actual realizations would be more complicated. Phase transition curves (in the space of two control variables) are represented by straight lines with slope of an appropriate sign, whereas they would actually be curves with slope of appropriate sign but with the magnitude of the slope varying continuously.

The new concepts that we introduced and tested across first order transitions in magnetic materials can be summarized as follows:

- The phase transition line in real materials is broadened into a band so that the transition is completed across a range of values of the control parameter (temperature or magnetic field). Other thermodynamic lines, specifically the limits of metastability (supercooling or superheating), are also broadened into bands.

- The phase transition can be arrested even under slow cooling if the temperature T_K at which such a kinetic arrest occurs falls above the limit of supercooling. If T_K overlaps the supercooling band at some value of the second control variable, then the phase transition can be interrupted and the transformation remains partial. Many interesting consequences like phase coexistence with continuously tunable phase fractions were actually observed.

- Since thermodynamic lines are broadened into bands, the same arguments imply that the T_K line is also broadened into a band. This also has interesting observable consequences.

The above concepts were introduced by us in our seminal paper [9] in 2001, and some predicted consequences were tested across a first order magnetic transition in doped $CeFe_2$. These were tested on other magnetic materials over the next few years but took many years to be accepted.

More consequences of these concepts were introduced and observed subsequently [12]. The evolution of concepts took place over a decade; they form the theme of this book. The presentation in this book will, however, not be chronological. The presentation benefits from hindsight and is pedagogical, introducing concepts one at a time.

The outline of this book is as follows. Chapters 1 and 2 cover the Ehrenfest classification and the modern classification, respectively. The emphasis is on first order phase transitions and on hysteresis and metastable states. In contrast with conventional texts, Chapter 1 provides a detailed discussion on the van der Waals model for the liquid–gas transition. It establishes that lines giving limits of metastability follow obviously in this model and discusses limitations of the Ehrenfest approach with regard to metastable states. Chapter 2 emphasizes metastable states in first order transitions, with a detailed discussion on the extent of metastability. It establishes that lines giving limits of metastability follow as a generic consequence for all first order transitions. It puts emphasis on the use of two control variables to check this extent of metastability and to cause transformation from a metastable to a stable state. These two chapters present conventional knowledge but set the background for the new material in subsequent chapters by establishing limits of metastability as thermodynamic lines similar to the phase transition line in the space of two control variables.

Chapter 3 starts with a discussion on the experimental results considered essential to establishing the existence of a first order phase transition but then shifts to a discussion on how hysteresis can be confirmed to have originated from metastable states around a first order phase transition. This represents the paradigm shift, away from latent heat and Clausius–Clapeyron relation, that has been used to classify many transitions in magnetic materials as first order.

Chapter 4 introduces the concept of arrest of the kinetics of transformation at a certain temperature (say T_K), a concept that has been addressed in the context of structural glasses for around a century. In contrast to structural glasses, we are considering this temperature as a function of the second control variable. We deviate from conventional knowledge, which presents this in the context of potential energy minima in real space, and discuss it in the context of the free energy as we discuss other transformation temperatures in the modern classification of phase transitions. This is because we shall look at the possible crossover between T_k and T_C as the second control variable is changed. This is a new concept we introduced in reference [9] and has been experimentally addressed in magnetic transitions because of the ease with which T and H can be varied *independently*. Many consequences of such variations, and their experimental observations, are brought out.

Chapter 5 introduces the additional concept of *broad* first order transitions. This, along with the ideas discussed in Chapter 4, was used in reference [9] to explain the intriguing observation of phase coexistence down to the lowest temperatures in half-doped manganites. These ideas predicted a continuously tunable phase coexistence at all (H, T) at the lowest temperatures, by using different cooling fields, that were then confirmed experimentally. These ideas also predicted a visually striking reentrant behavior on CHUF, with particular inequalities in these field magnitudes, which was again verified experimentally. These new developments are presented in Chapter 5. There are many magnetic materials showing phase transitions with features discussed in this chapter. Disorder-induced broadening of transitions and the overlap of kinetic arrest with the limit of supercooling appears to be widespread. We can expect many applications of these concepts in functional magnetic materials.

"Why do glasses form?" is a question that has persisted for a very long time. There has been a folklore that multiatom materials or long-chain compounds form glasses easily because of entanglement during positional ordering required for crystallization. About half a century ago, metallic glasses were formed by rapid cooling, supporting the belief that lowering diffusivity is essential to arresting structural reconstruction. Extensive studies on magnetic materials have established kinetic arrest of first order transitions that do not involve diffusion of atoms. Viewing this kinetic arrest as a generalization of the formation of a structural glass, the cause for this arrest should provide a more general answer to "why do glasses form?" This will be addressed at the end of this book.

References

1. R.M. White and T.H. Geballe, *Long Range Order in Solids*, Academic Press, New York (1979), p. 12.
2. A. Schilling, R.A. Fisher, N.E. Phillips, U. Welp, D. Dasgupta, W.K. Kwok, and G.W. Crabtree, *Nature* **382** (1996) 791.
3. S.B. Roy and P. Chaddah, *J Phys Cond Matt* **9** (1997) L625.
4. S.B. Roy, P. Chaddah, and L.E. DeLong, *Physica C* **304** (1998) 43.
5. S.B. Roy, P. Chaddah, and S. Chaudhary, *Phys Rev B* **62** (2000) 9191.
6. H. Kuwahara, Y. Tomioka, A. Asamitsu, Y. Moritomo, and Y. Tokura, *Science* **270** (1995) 961.
7. Y. Imry and M. Wortis, *Phys Rev B* **19** (1979) 3580.
8. M. Fäth, S. Freisem, A.A. Menovsky, Y. Tomioka, J. Aarts, and J.A. Mydosh, *Science* **285** (1999) 1540.
9. M.A. Manekar, S. Chaudhary, M.K. Chattopadhyay, K.J. Singh, S.B. Roy, and P. Chaddah, *Phys Rev B* **64** (2001) 104416.

10. A. Banerjee, A.K. Pramanik, K. Kumar, and P. Chaddah, *J Phys Cond Matt* **18** (2006) L605.
11. A. Banerjee, K. Mukherjee, K. Kumar, and P. Chaddah, *Phys Rev B* **74** (2006) 224445.
12. A. Banerjee, K. Kumar, and P. Chaddah, *J Phys Cond Matt* **21** (2009) 026002.

Acknowledgments

This book is based on new ideas that were introduced as we attempted to understand anomalies that were being reported in measurements on magnetic materials that indicated first order transitions as magnetic field (H) or temperature were varied. These materials were then known as the "giant magnetoresistance (GMR) manganites." After having studied unusual behavior across first order transitions in vortex matter for about five years, we were eager to try our unusual ideas on the GMR materials. But the field was very active due to anticipated potential for applications, and some of us had experienced similar activity in the high-T_C superconductors. Given our comparative physical isolation, Dr. Sindhunil Roy and I decided to stay a bit away from the strong tide of GMR research and try out our ideas on a material of "academic" interest. Dr. Roy had done extensive work on $CeFe_2$ that showed a low temperature ferromagnet-to-antiferromagnet transition (only) under various dopings. This transition was accompanied by a sharp increase in resistance that was, however, much smaller than that seen on charge-ordering in the half-doped GMR materials. Sindhunil emphasized the similarities and the fact that the smaller magnetoresistance would keep this material in the "of academic interest" category. We, along with younger colleagues, took the plunge with Al-doped $CeFe_2$ and published data supporting our unusual ideas in 2001. We could not publish in the most prestigious journals, but *Physical Review B* was an archival journal that was highly respected and well read. We continued testing our ideas with newer experiments and were regularly publishing, drawing attention to similarities with GMR manganites, and were not unhappy that we were left alone to our new thoughts. It was a few years before our work was referred to in studies on newer materials, by Magen et al. and by Zhang et al. I record my gratitude to Sindhunil Roy.

Around this time, I switched institutions and had an administrative role over a group that was doing active research in these GMR manganite materials. I was loath to collaborate and test unusual ideas under such administrative relations but was happy to find that Dr. Alok Banerjee had no hesitation in challenging the unusual ideas we proposed. He had the experimental wherewithal, and the attitude, to conduct new failure-test experiments with GMR manganites. We had a wonderful collaboration and were soon joined by Dr. Rajeev Rawat and others.

Given that many of our new ideas were presented in Manekar et al. in *Physical Review B* in 2001, why has this book taken so long to write? A massive experimental effort was essential to establish our ideas and to create new experimental protocols and generate failure-test data that could not be explained by what I would now call alternate models. Further, I do not believe in prepublication validation; validation is done by postpublication review by identifiable experts. The ideas in this book have survived that validation.

I now list the many coauthors with whom I have had personal interactions and who have helped me learn. I thank Dr. Meghmalhar Manekar, Dr. Maulindu Chattopadhyay, Prof. Sujeet Chaudhary, and Dr. Kanwaljeet Singh Sokhey, who were close colleagues at the erstwhile Centre for Advanced Technology. At the UGC-DAE Consortium for Scientific Research, I had the advantage of working with colleagues as well as with many who were then students. I thank Dr. Vasudev Siruguri, Dr. Asim Pramanik, Dr. Kaustav Mukherjee, Dr. Kranti Kumar, Dr. Suryanarayan Dash, Dr. Archana Lakhani, Dr. Pallavi Kushwaha, Dr. Vasant Sathe, Dr. D. Kumar, and Dr. Pallab Bag. Dr. Siruguri and his colleagues provided very convincing neutron diffraction data under the new CHUF protocol. Prof. R. V. Ramanujan and Prof. V. K. Pecharsky were collaborators from abroad, whom I thank for making an impact beyond what can be acknowledged by joint authorship.

A person who likes solitude, probably necessary for evolving unusual explanations, is sometimes also termed as a bit "difficult." My *ardhangini** of nearly 40 years has always supported me and has not let me feel my flaws. Thank you, Kumud!

Finally, I must thank CRC Press and, particularly, Aastha Sharma for goading me into writing this book.

---

* *Ardhangini* is a Hindi word for wife, implying that you exist together, and would be halved without her.

Author

Praveen Chaddah, PhD, joined Bhabha Atomic Research Centre (BARC) through their Training School in 1973. His PhD work involved setting up a Compton Profile Spectrometer with a γ-ray source (the first in India) for electron momentum density measurements. His work focused on electron states in structurally disordered materials and on electron correlation effects. As a postdoc at the University of Illinois at Urbana-Champaign, he initiated measurements of nuclear momentum densities in the quantum solid 4He using the spallation neutron source IPNS at Argonne. Dr. Chaddah also worked on correlating the superconducting and martensitic transitions in A-15 superconductors. He worked on the development of superconducting magnets and of multifilamentary NbTi wires at BARC during 1982–1987 and later made important contributions to the development and extension of Bean's Critical State model for the high T_C superconductors. He reformulated this as a "minimum flux-change" hypothesis and contributed to its application to sample shapes having finite demagnetization factor. He then worked on first order phase transitions in vortex matter in superconductors, as also in magnetic materials. His recent emphasis has been on understanding metastabilities associated with supercooling and superheating, as also those associated with glass-like arrest of kinetics. His work showing tunability of coexisting phases in half-doped manganites by varying the cooling field was followed up on many materials exhibiting magnetic-field-induced first order transitions. This introduced the idea of "kinetic arrest" as a broad first order transition that is interrupted before completion. He developed the protocol CHUF (cooling and heating in unequal fields), which has provided rather visual evidence of kinetic arrest resulting in a glass-like arrested state. The observation of such behavior across magnetic first order transitions in various magnetic materials, where diffusive motions are not apparent, may lead to newer understanding of what causes glasses to form. Dr. Chaddah received the INSA Young Scientist medal in 1978 and the MRSI-ICSC Prize for Superconductivity in 1993. He is a fellow of the Indian National Science Academy, the Indian Academy of Sciences, and the National Academy of Sciences of India. He is also an elected member of the Asia Pacific Academy of Materials. He has authored more than 200 research papers.

Dr. Chaddah was the director of the UGC-DAE Consortium for Scientific Research for more than eight years, where he established internationally competitive facilities for experimental research that were open to researchers from universities across India. He retired from the Department of Atomic Energy in December 2013.

1

Phase Transitions, and Rigorous Definitions

1.1 Phase Transitions in Nature, and a Meandering Introduction

The term "phase transition" is often used colloquially with the criterion that a phase transition occurs when incremental changes cause a sudden qualitative change in properties, instead of just having incremental effects. The most commonly observed phase transition is the melting of ice. The neuroscientist Ramachandran [1] describes this very well for a nonspecialist through the gradual warming up of a block of ice, noting that with an incremental change in temperature well below the melting point "all you have that you didn't have a minute ago is a slightly warmer block of ice," until one reaches the critical melting temperature when "incremental changes stopped having incremental effects, and precipitated a sudden qualitative change called a phase transition."

The change from ice to water, or from water to steam, is an example of incremental change—which was causing an incremental effect—suddenly causing a striking effect. Ramachandran then suggests that incremental changes in the size of the brains of our ancestors over millions of years followed gene-based Darwinian evolution, but then culminated in a "mental phase transition" about 150,000 years ago. According to him, the incremental increase in the size of the brain resulted in succeeding generations of apes having slightly better motor and mental abilities, and the hominid brain reached nearly its present size about 300,000 years ago. He argues that this mental phase transition was associated with the development of certain small but key brain structures so that "old parts were there, but they started working together in new ways that were far more than the sum of their parts. This transition brought us things like full-fledged human language, artistic and religious sensibilities, and consciousness and self-awareness" resulting in the ability to wonder. According to him, these uniquely human attributes, which distinguish us from lower primates, came as a result of a phase transition. Such an exotic usage of the term "phase transition" is frequently encountered in discussions on social or financial upheavals that occur on short time scales. This definition of incremental changes causing drastic effects is, however, not acceptable to physicists because we like to talk about causing changes in a parameter that can be controlled (called "control parameter") and observing effects in properties that can be measured and quantified (called "physical properties"). It should be possible to vary the control parameter in an actual experiment or in a theoretical calculation, and Ramachandran's interesting hypothesis of a mental phase transition in the evolution to human form could have been tested (albeit through model simulations) if we had a solvable model for brain activity in which physical linkages in the brain could be identified as variable control parameters.

The idea that phase transitions can and do occur in the human brain, as it exists today, has also been pursued at more computable and measurable levels. Changes such as from being awake to sleep and from slow-sleep to REM sleep are being discussed as showing features of first order phase transitions [2]. We recognize, as physicists, that phase transitions are thermodynamic phenomena that are talked about with rigor in various branches of science. Phase transitions in the early universe are discussed, for example, to explain the observed particle–antiparticle asymmetry, to explain a remnant magnetic field, and as a source of gravitational waves, even though control parameters in these cases can only be varied in model calculations and not by human experimenters.

Stated more rigorously, phase transitions occur when a thermodynamic system changes from one phase or state of matter to another. A system changes phase when some measurable physical property changes sharply under a small change of a control parameter (there is a "striking effect"), the most common control parameter being temperature. Within a phase, a similar or larger change of the control parameter will only cause small changes in the same property. As an example, in the most common phase transition, the volume of ice changes by about 10% when it melts into water, with the melting process occurring with no measurable change of temperature. We have been taught not to put a bottle full of water into the freezer because the sharp increase of 10% in volume, as the water freezes, can cause the bottle to explode. This change in volume is definitely drastic. At lower or higher temperatures, temperature changes of even a few degrees cause only comparatively slight changes in volume that are perceptible only with sensitive measurements. To sum up, some physical property (volume, in this case) will change discontinuously (by about 10%, in this case) with no change in temperature when there is a phase transition.

Phase transitions, consistent with this more rigorous usage of the term, are ubiquitous in nature. The most common phase changes are between liquid and solid phases and between liquid and gas phases. Our everyday experiences of phase transitions include the melting of ice into water and the boiling of water into steam. The melting of glaciers—as distinct from a rise in their temperature while still frozen—is a first order phase transition requiring input of large amounts of heat. As a natural phenomenon, this is irreversible in that water can flow away from where the glaciers are located. This is continuously discussed and evaluated by environmentalists in the context of global warming and the potential rise in ocean levels that could threaten life on our planet.

The melting of ice into water is also an everyday living experience, and it is not really irreversible when in a confined space. It is common knowledge that the temperature of a glass containing a mixture of water and ice does not rise as long as some ice is present. The total volume of ice and water decreases as more and more ice melts, consequent to the sharp reduction in volume in this phase transition; however, the process is slow and may not be immediately obvious in a living room because some of the liquid water is also being consumed. What is obvious is that both ice and water coexist, even as the fraction of ice in the glass reduces with time. This does tell us that the ice–water mixture can absorb heat without a temperature rise, signifying a "latent" heat (this term was coined to distinguish it from the usual or "sensible" heat that raises temperature [3]). Phase coexistence and latent heat are characteristics of this type of phase transition.

The conversion of water into steam is another everyday experience. The most obvious feature of this phase transition is the large orders of magnitude change in volume. This drastic change in volume is an obvious characteristic of a liquid-to-gas phase transition. As noted earlier, solid-to-liquid phase transitions are also accompanied by sudden changes in volume, but the magnitude of change is smaller in these cases. (Also, there are no known examples of the volume of a liquid decreasing on transforming to the gas phase, while there are many examples of the volume of a solid decreasing on transforming to the liquid phase.) The visual observation of the huge volume increase in the liquid–gas phase transition led to the steam engine and, subsequently, to the Industrial Revolution—one really path-breaking application of phase transition.

Solid-to-liquid and liquid-to-gas transitions are observed in various materials on heating, and typical phase diagrams are shown in Figure 1.1. Both these transitions involve absorption of latent heat. As mentioned earlier, the temperature of a solid material will rise continuously on steady heating but will become constant as the solid starts melting. Once the entire solid has melted, then and only then will the temperature (now of the liquid) start rising again. The temperature will once again become constant when the liquid starts boiling into a gas. Once the entire liquid is converted to gas, then and only then will the temperature of the gas start rising. This is depicted in the schematic in Figure 1.2. The rate of the rise in temperature within a phase is slower for a phase with higher specific heat.

The reverse gas-to-liquid and liquid-to-solid transitions are observed on steady cooling, with the temperature again becoming constant when the gas is converting to liquid and also when the liquid is converting to solid. This temperature vs. time profile is used as a signature of the temperature at

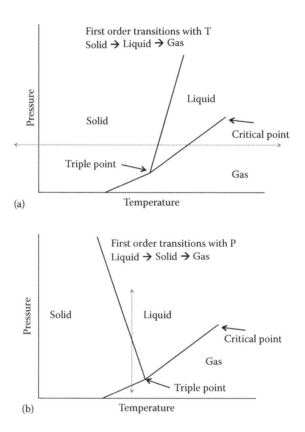

FIGURE 1.1
We show two schematic solid–liquid–gas phase diagrams, corresponding to (a) the more common case of the liquid having a lower density than that of the solid and (b) the less common but frequently encountered case of the liquid having a higher density than the solid. The order in which the three phases are encountered with varying temperature is the same in both cases, but the order encountered with varying pressure is different. In (a), the liquid transforms to solid with increasing pressure while in (b) the solid transforms to liquid. This is because increasing pressure always causes a transformation to the phase with higher density. We emphasize that the various phases are separated by phase transition lines that actually change slopes continuously. We, however, assume in these and in most schematics in this book, for simplicity and without affecting the arguments that introduce new ideas, that these phase transition lines have a fixed slope.

which first order transition takes place and is normally considered reversible between the heating and cooling cycles.

When plotted as a function of temperature, the melting of a solid is signaled by a sudden change in volume, as depicted in Figure 1.3. In most cases this is a sharp increase, but in many cases it is also a sharp decrease with the change in volume being of the order of a few percent. The conversion of a liquid to gas is accompanied by many orders of magnitude increase in volume, with no known case of a decrease in volume.

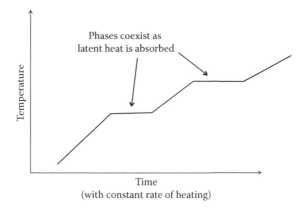

FIGURE 1.2
We show the schematic temperature vs. time profile for the solid → liquid → gas transition, with a constant heat input. The same profile is valid for both cases of Figure 1.1. The specific heat in this schematic is assumed to be higher in the higher temperature phase, though this is not necessary and may not always be the case.

It is a widely held belief that indefinite heating of every solid will convert it to liquid (or gas) and lowering the temperature will result in the liquid converting back to solid. However, this cannot be stated as an axiom because there is one known counterexample. This interesting exception is the liquid–solid transition observed in helium-3 that shows anomalous behavior on heating and cooling in a certain range of pressure around 30 atmospheres. Its phase diagram is shown in Figure 1.3c. The liquid form of this isotope of helium freezes on heating, and further heating causes it to melt again and eventually convert to gas. The solid at about 0.3 K and 30 atmospheres pressure does melt on cooling. Since even one counterexample is enough to disprove an axiom, we must discuss this one material in some detail before we can assert any universal axiom on the role of increasing temperature in causing phase transitions. We shall do so shortly.

Returning to everyday experiences, we now discuss the role of pressure in causing a phase transition. As mentioned, water boiling into steam is an everyday experience in the kitchen. Pressure cookers are very common in India as a means for reducing energy consumption. The weight on a pressure cooker maintains water at about 20°C above its normal boiling point, and this reduces the cooking time because food cooks faster when it is maintained at a higher temperature. Also, when water is maintained at 100°C, some of it converts into steam and the latent heat for boiling—which is about 25 times the heat required to heat water from 100°C to 120°C—is lost in the steam that in no way aids cooking, representing lost energy. If the weight is removed suddenly, then the pressure falls and water at 120°C converts to steam with a large increase (by a factor of over 1000) in volume. Every person is taught not to remove the weight while hot, because this sudden drop

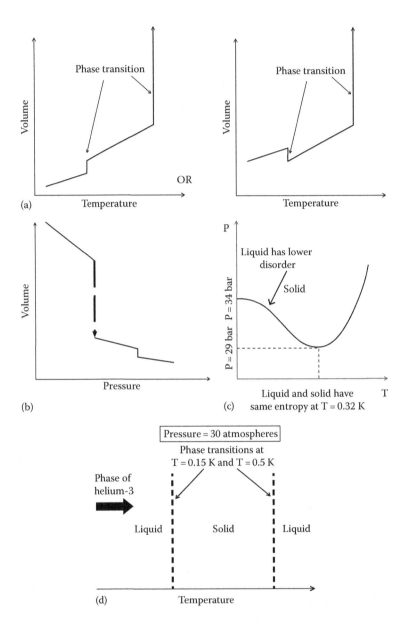

FIGURE 1.3
(a) We show the schematic volume vs. temperature profile for the solid–liquid–gas phase transitions. The volume may rise (or fall) at the solid–liquid transition but rises sharply at the liquid–gas transition. (b) We show the schematic volume vs. pressure profile encompassing the solid–liquid–gas phase transitions. The highest-pressure phase could be the liquid or the solid phase, while the lowest-pressure phase is the gaseous phase. In (c), we show the phase diagram of helium-3 as a function of temperature and pressure, and in (d) we show the phase transitions in helium-3 as a function of temperature at a pressure of about 30 atmospheres.

in pressure at a temperature above 100°C releases large volume of hot steam that can cause serious burn injury. The weight should not be removed—or the pressure experienced by hot water should not be reduced—until the water has cooled down. We thus recognize that reduction in pressure at a suitable fixed temperature does cause transition to the phase with lower density. Similarly, increase in pressure at a suitable temperature does cause a transition to the phase with higher density. Our common experience, again based on water, is of ice melting into the higher-density liquid phase under application of pressure. Let us take two examples. Ice-skating is based on ice melting under pressure as the skater's weight is transmitted through the blades under the skates. There is another common experience in some parts of the world where ice-slabs are cut by using a thin metal string loaded at both ends with small weights—the pressure melts ice under the string and the slab separates into two parts as the string falls through. These are common examples that pressure change, too, can cause a phase transition. The common theme here is that increasing pressure causes a transition to the phase with higher density and decreasing pressure causes a transition to the phase with lower density. This generic surmise on phase transitions under pressure is actually valid in all phase transitions known—it forms an axiom and can also be understood as another example of the widely applicable Le Chatelier's principle. The principle states in its popular usage that when any system at equilibrium is subjected to an external change, then the system readjusts itself to (partially) counteract the effect of the applied change. Since compressibility is positive in all physical situations, increased pressure always results in a reduced volume to counteract the increased pressure. There is, however, no such obvious relation to dictate that melting counteracts a rise in temperature. Le Chatelier's principle does not guide us as we still seek a universal axiom on the role of increasing temperature in causing phase transitions.

Phase transitions can also be caused by changes in the magnetic field. This happens in magnetic materials, which are the topic of this book. The superconducting transition can be characterized as a magnetic transition because diamagnetism develops when superconductivity sets in. The superconducting transition is depicted in Figure 1.4. It also has some special features. The physical property that changes drastically, and for which the superconducting state has many applications, is the electrical resistance, which drops to zero as temperature is lowered. This zero-resistance state, as depicted in the schematic in Figure 1.4, can be transformed back to the normal state by either heating or applying a large enough magnetic field. For physicists, resistance is a physical property but not an equilibrium property, and we shall be addressing this difference repeatedly in subsequent chapters. The equilibrium property that changes in this transition is magnetization, which reduces suddenly at the superconducting transition. This change in magnetization was found by Meissner and Oschenfeld [4] in 1933, many years after the drop in resistance was observed by Onnes [5] in 1911. This simultaneous

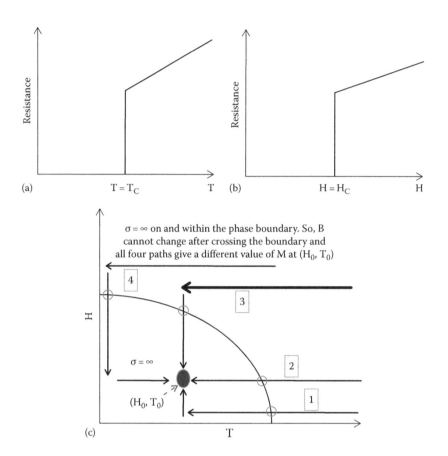

FIGURE 1.4
We show schematics for the resistance vs. temperature behavior in (a) and the similar resistance vs. magnetic field behavior in (b) for a typical superconducting phase transition. In (c), the schematic phase diagram shows that the superconducting state undergoes a transition to the normal state both with increasing temperature and increasing magnetic field. This is consistent with the superconducting state being diamagnetic and the normal state having higher magnetization, with the magnetic field being totally, or partially, expelled when the transition to the superconducting state takes place. We show four different paths of reaching the point (H_0, T_0) where the sample is superconducting. The state at (H_0, T_0) was believed to be path-dependent until the Meissner effect was established by Meissner and Oschenfeld [4].

change in magnetization explained that the superconducting transition must also be influenced by a magnetic field H. An increasing H in the superconducting state causes a transition to the normal state, which has higher magnetization. This feature—of increasing H causing a transition to the phase with higher magnetization—is a general feature of all magnetic transitions, with no known exceptions. Just as a higher pressure favors a state with smaller volume or higher density, a higher magnetic field favors a state with higher magnetization. This can also be understood as another example of Le Chatelier's principle. However, there is no general axiom on how electrical resistance changes when a phase transition is caused by an increasing magnetic field, and the statement made for superconductors is not applicable to other transitions signaled by a sharp change in resistance. Increasing the magnetic field does not always cause a transition to the state with higher resistance.

We come back to the question "is there one property that changes in a universal way when a phase transition is caused by increasing temperature?" Increasing the temperature of any material does cause increased oscillations of the constituent atoms or molecules, and there is a belief (known as Lindemann's criterion) that increasing oscillations do cause a solid to melt; but we have given a counterexample, as depicted in Figure 1.3c. We argue in the following text that the general axiom we are looking for can be stated as "increasing temperature causes a transition to the phase with higher disorder." In thermodynamics, disorder is quantified by entropy (see, e.g., [7]), though entropy is measured experimentally only through its derivative, i.e., in specific heat measurements. Entropy has contributions from all phase space, and disorder is not solely about structural disorder. All liquids are more disordered than their corresponding solid phases in terms of their positions in real space, and this governs the common observation that solids melt on heating. Nevertheless, liquid helium-3 is a quantum Fermi liquid and is highly ordered in momentum space. The entropy of the liquid varies approximately linearly with T as $[(T/T_F^*)\ln2]$, where T_F^* is the effective Fermi temperature. On the other hand, the spins in solid helium-3 are not aligned until it is cooled to about 1 mK, and its entropy is approximately constant at $R\ln2$ above about 10 mK. Both T_F and the magnetic interaction between nuclear spins depend on the interatomic distances and thus on pressure. In a certain pressure range around 30 atmospheres, the solid has higher entropy till about 320 mK. Heating lowers the free energy of the phase with higher entropy, and the liquid makes a transition to solid on heating. The temperature at which this freezing on heating takes place drops as pressure rises because the entropy of the solid reduces in comparison to that of the liquid, with this temperature dropping to 0 K at about 34 atmospheres. At higher pressures, solids have lower entropy than liquids, and is the equilibrium phase or the ground state.

The statement "increasing the temperature causes a transition to the phase with higher entropy" has no known exception. Entropy rises discontinuously at the transition temperature for a first order phase transition. Entropy depends on the total disorder and not just on structural disorder. We have to remember this in the magnetic first order transitions between ferromagnetic (FM) and antiferromagnetic (AFM) phases, where there is no universality about which phase has higher entropy. The earlier discussion has emphasized that the inequality in entropy can be counterintuitive. We have discussed the example of helium-3, where the liquid has lower entropy than that of the solid in a certain (H, T) region. We shall encounter phase transitions in magnetic materials where the AFM phase is more ordered than the FM phase, in addition to the more intuitive case of the FM phase being more ordered than the AFM phase.

1.2 Introduction: Back to Basics

We now come back from these heuristic arguments to our knowledge of elementary thermodynamics [7]. For a system kept at constant temperature and pressure, the equilibrium state is one that minimizes the following:

$$G = U + PV - TS \tag{1.1}$$

where the Gibbs' free energy G, the internal energy of the system U, the volume V, and the entropy S are all extensive variables. We have also discussed magnetic materials whose magnetization M (also an extensive variable) depends on the applied magnetic field H, and G must now be chosen as follows:

$$G = U + PV - TS - MH \tag{1.2}$$

We depict in Figure 1.5 the role of three control variables—T, P, and H—in evolving G through the corresponding physical properties—entropy, volume, and magnetization. We can consider phase transitions caused by other control variables in a similar manner, and each such control variable must influence the free energy of the system through its product with the corresponding physical property. In case we use any other control variable (like electric field for a dielectric material that could undergo a paraelectric to ferroelectric transition), an appropriate additional term would be added to G. Since G must be minimized in the equilibrium state, it follows obviously that if the only variable control parameter is T, then with an increasing T the phase transition would be to the phase with higher entropy. Similarly, if the only variable control parameter is H, then with increasing H the phase transition would be to the phase with higher magnetization. Further, if the only variable control parameter is P, then with increasing P the phase transition would be to the phase with lower V or higher density. This is depicted in the schematic in Figure 1.5.

From this very general discussion on "phase transitions" observed in nature and in everyday experiences, we now define phase transition. During phase transition, some physical properties of the medium should change discontinuously. The change in resistance in a normal-superconducting transition and the change in volume in a liquid–gas transition provide examples of such discontinuous changes that are drastic. A continuous (even nonmonotonic)

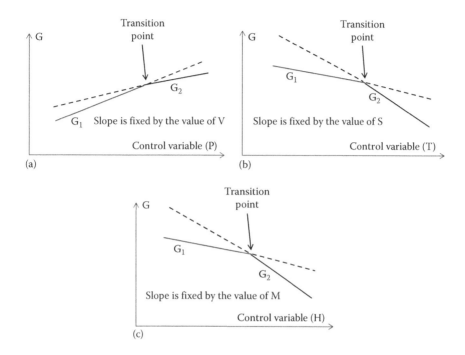

FIGURE 1.5
We consider two phases with sharply different values of specific volume V, entropy S, and magnetization M. We ignore the comparatively small variations (within each phase) with pressure, temperature, and magnetic field. Assuming that we vary only one of these control variables and keep the other two fixed, the free energies G_1 and G_2 have a fixed slope as we vary (a) only pressure, (b) only temperature, and (c) only magnetic field. The control variable where G_1 and G_2 cross corresponds to the system making a transition from one phase to the other; the phase with the lower value of G is shown by the full line and corresponds to the equilibrium phase.

variation like the change in density as water is heated from 1°C to 99°C does not imply a phase transition. Put with mathematical rigor, we note that some physical property must change in a discontinuous way at a phase transition. Most physical properties are the derivatives of free energy, and we thus expect some derivative of G to be discontinuous at a phase transition. G must, then, be a nonanalytic function of the control variable at the value of the control variable where the phase transition occurs. The schematic in Figure 1.5 depicts a discontinuity in the first derivative, and dashed lines show a linear extrapolation in the region beyond the phase transition point.

The existence of nonanalyticity in G as a function of a control variable implies a phase transition. To understand this, let us consider two possible phases (phase-1 and phase-2) of a material at a given value of the control variables. Let us assume that the properties that are different are first derivatives of G with reference to a control variable. We label the properties in the two phases by subscripts 1 and 2. For simplicity, we assume that these properties are very different in the two phases but are only weakly dependent on the control variables within each phase—or that we can ignore the dependence of the physical property (within a phase) on the control variable. Under this assumption, the free energies in the two phases would vary linearly with each control variable, and we get the following:

$$G_1 = U_1 + PV_1 - TS_1 - M_1H$$

and

$$G_2 = U_2 + PV_2 - TS_2 - M_2H \qquad (1.3)$$

The phase that exists at a given (P, T, H) would be the one with lower free energy. Let us assume that $S_1 < S_2$, and we compare $G_2 - G_1$ at (P, T, H) and at (P, T + ΔT, H), assuming that we vary only T, keeping P and H fixed. We note that since S is positive, G will always fall with rising T. In the first approximation, we ignore the T-dependence of U, S, V, and M. ($G_2 - G_1$) would now change by the following:

$$\Delta(G_2 - G_1) = \left[G_2(P,T+\Delta T,H) - G_1(P,T+\Delta T,H) \right]$$
$$- \left[G_2(P,T,H) - G_1(P,T,H) \right] = -(S_2 - S_1)\Delta T \qquad (1.4)$$

Since $S_2 > S_1$, this would decrease as ΔT rises, with G_2 eventually becoming equal to and then smaller than G_1. This is depicted schematically in Figure 1.5b. The dashed line represents the phase with higher free energy, indicating that it is not the equilibrium phase. If the system can exist in this higher free energy phase, then it must be metastable and should transform (or relax) to the stable phase corresponding to the solid line. These metastable phases are referred to as supercooled or superheated states. No general predictions can be made based on the discussion so far as to whether such metastable phases can exist, and whether there is any limiting temperature for the existence of such metastable states. The temperature T_C at which G_1 becomes equal to G_2 is the temperature at which a phase transition, from phase-1 to phase-2, will occur under equilibrium. The equilibrium free energy is

$$G_{Eq}(T) = G_1(T), \quad \text{for } T < T_C$$
$$= G_2(T), \quad \text{for } T > T_C$$

G_1 and G_2 have different dependence on T, with slopes given by S_1 and S_2. $G_{Eq}(T)$ changes slope from S_1 to S_2 at T_C. Equation 1.4 can be written as

$$\Delta(G_2 - G_1)/\Delta T = -(S_2 - S_1)$$

and in the limit $\Delta T \to 0$ we get

$$\partial(G_2 - G_1)/\partial T = -(S_2 - S_1)$$

or

$$\partial \Delta G / \partial T = -\Delta S \tag{1.5a}$$

Following similar arguments, if we vary only the pressure, then $G_{Eq}(P)$ changes slope from V_1 to V_2 at P_C, and if we vary only the magnetic field, then $G_{Eq}(H)$ changes slope from M_1 to M_2 at H_C. We similarly obtain the relations

$$\partial \Delta G / \partial P = \Delta V \tag{1.5b}$$

$$\partial \Delta G / \partial H = -\Delta M \tag{1.5c}$$

We can also have a phase transition in which entropy, density, and magnetization are the same in the two phases and each of ΔS, ΔV, and ΔM is zero, but their derivatives and the corresponding physical properties (specific heat, compressibility, and susceptibility, respectively) are different. The transition again occurs at the value of the control variable where the two free energies cross, as depicted in the schematics in Figure 1.6. As is apparent, the discontinuity in G at the phase transition gets more and more subtle as the derivative

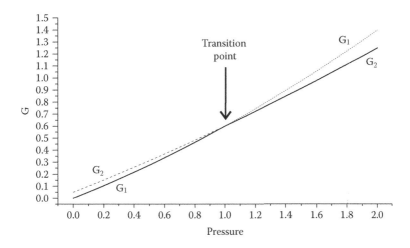

FIGURE 1.6
We depict a situation where the first derivatives of G are continuous at the phase transition. The values of V, S, and M are not very different in the two phases and are actually equal at the phase transition point, but their variations with the control variables are different. G_1 and G_2 cross as the control variable (P in this schematic) is changed, but the crossing is smooth compared to that depicted in Figure 1.5.

showing the discontinuity becomes of higher order. The physical property corresponding to higher derivatives of G is more difficult to measure, and the discontinuity becomes even more difficult to establish. We now enunciate the rigorous Ehrenfest classification of phase transitions, which accounts for all such possibilities.

1.3 Ehrenfest Classification

Physical properties, specifically, equilibrium properties, are derivatives of free energy. At phase transition, a physical property would change discontinuously, so it follows that some derivative of free energy should change discontinuously at phase transition. Equilibrium properties such as entropy, density, and magnetization are the first derivatives of G with respect to control variables. A discontinuous change in these would be experimentally obvious, and the phase transition would be drastic, as shown in Figure 1.5. Second derivatives are properties such as specific heat, compressibility, and susceptibility. These properties are comparatively more difficult to measure, and so are discontinuous changes in these. The phase transition, thus, would be more subtle. The third (and higher) order derivatives of free energy are properties that are measured even less frequently. Thus, discontinuities in derivatives of an increasing order are experimentally less obvious.

This idea underlies the Ehrenfest classification of phase transitions. The order of a phase transition is defined as that of the lowest derivative of the free energy that is discontinuous at the transition. (The derivative of one order higher is then not defined, and the derivative of one order lower has to be continuous.) Let us follow these definitions of the Ehrenfest classification.

In "first order" phase transitions, G varies continuously as a control variable is varied (it always varies continuously, whether or not there is a phase transition) whereas its first derivatives have to be discontinuous at the transition. First derivatives of free energy are entropy, volume, and magnetization.

$$S = -\left(\partial G/\partial T\right)_P$$

$$V = \left(\partial G/\partial P\right)_T$$

$$M = -\left(\partial G/\partial H\right)_T \tag{1.6}$$

Their discontinuities at the phase transition were related in Equations 1.5a through 1.5c to the derivatives of $\Delta G = (G_1 - G_2)$ at the transition point. A discontinuity in volume or magnetization is measured directly, whereas that in entropy is observed as latent heat or infinite specific heat. As we derive in the following, latent heat is related to the discontinuity in volume or magnetization through the Clausius–Clapeyron relation. We have the cyclic chain identity as follows:

$$\left(\partial P/\partial T\right)_{\Delta G}\left(\partial T/\partial\Delta G\right)_{P}\left(\partial\Delta G/\partial P\right)_{T}=-1$$

Or, using Equations 1.5a through 1.5c, we have the following:

$$\left(\partial P/\partial T\right)_{\Delta G}\left(-\Delta S\right)^{-1}\Delta V=-1 \tag{1.7}$$

On the first order phase transition line, where $\Delta G = 0$, we get the Clausius–Clapeyron relation

$$dP/dT = \Delta S/\Delta V = L/(T\Delta V) \tag{1.8}$$

We can similarly derive for the slope of the phase transition line in the space of H and T and relate it to the latent heat and magnetization jump at that transition point. Satisfying this Clausius–Clapeyron relation has historically been considered a necessary and sufficient condition for classifying a phase transition as first order. The Ehrenfest classification is mathematically rigorous.

In "second order" phase transitions, entropy, volume, and magnetization vary continuously, whereas the second order derivatives of G are discontinuous at phase transition. With this definition, a second order transition is more subtle in that there is no volume change and no latent heat. Discontinuities are observed in specific heat, compressibility, and susceptibility, which are related to the free energy as follows:

$$C_P = T\left(\partial S/\partial T\right)_P = -T\left(\partial^2 G/\partial T^2\right)_P$$

$$\beta = -1/V\left(\partial V/\partial P\right)_T = -1/V\left(\partial^2 G/\partial P^2\right)_T$$

$$\chi = -\left(\partial^2 G/\partial H^2\right)_T \tag{1.9}$$

Under the Ehrenfest scheme, phase transitions of arbitrary integral order could exist. For a phase transition of nth order, the nth derivative of G must be discontinuous, the $(n-1)$st derivative must be continuous, and the $(n+1)$st derivative cannot be defined. The classification is mathematically rigorous and appealing but loses physical significance for increasing order. As mentioned earlier, the physical properties corresponding to higher order derivatives of G are not easily measured, and the corresponding physical difference between the two phases may not be apparent. There are no phase transitions known yet in which the third or higher derivatives of G are discontinuous, though at least one claim of a fourth order transition has been published [6].

Further, and more importantly, there are situations in which a derivative of some order is continuous but singular, and the singularity ensures that the derivative of the next higher order does not exist. So we have nonanalytic behavior in G, but there is no derivative of integer order that is discontinuous. Not only is this a mathematical possibility, such behavior in the specific heat diverging logarithmically has been observed in various cases [7]. Consider the case

$$C_P = -T\left(\partial^2 G / \partial T^2\right)_P = -A\log\left(T - T_C\right) \quad \text{for } T > T_C$$

and

$$C_P = -T\left(\partial^2 G / \partial T^2\right)_P = -A\log\left(T_C - T\right) \quad \text{for } T < T_C \qquad (1.10)$$

Then, the second derivative of the free energy C_P is continuous and the order of the transition cannot be 2 and must be higher; however, the next higher derivative is not defined at T_C since it is of magnitude infinity but with slopes of opposite sign as T_C is approached from above or below. The order of the transition cannot be 3 since the third derivative is not discontinuous—it just does not exist. This transition does not fit into the Ehrenfest scheme. Such logarithmically diverging specific heat—actually observed in some transitions—has often been cited [7] as a major limitation of the Ehrenfest classification. It is now considered to be an incomplete method for classifying phase transitions that may be higher than the first order, but there appear to be no issues with classification of first order transitions by this method.

Since this book is about first order transitions, we shall now look at the limitations of the Ehrenfest scheme beyond just the classification issues. We shall address in some detail its conceptual limitations in describing metastable states. We shall first discuss in detail an exactly solvable model of a first order transition that makes predictions on such metastable states. We shall identify questions in the context of first order transitions that remain unaddressed in the Ehrenfest classification and have been raised recently in the context of the first order magnetic transitions in real materials with potential for applications.

1.3.1 First Order Transitions

As mentioned earlier, latent heat exists only for first order transitions. Consequently, relating latent heat to discontinuities in derivatives of the free energy, as is done through the Clausius–Clapeyron relation, is also specific to these first order transitions. Checking the Clausius–Clapeyron equation given by Equation 1.8 is demanding, because it needs to be satisfied all along the phase transition line in the two-control-parameter space. It can also be experimentally tough (e.g., when the latent heat is small), and it sometimes also becomes difficult even to distinguish latent heat (or infinite specific heat) from a large peak in specific heat [8]. We shall illustrate these difficulties with an example from a phase transition in superconductors.

The superconductor-to-normal transition for a type-I superconductor is first order in the presence of a magnetic field [9]. In type-II superconductors, we do not have perfect diamagnetism for $H > H_{C1}$ and flux penetrates in the form of infinitely long vortices parallel to the external magnetic field, each vortex carrying a quantum of flux $\varphi_0 = hc/2e$. The dimension of the vortex perpendicular to the field increases continuously as the superconducting-to-normal transition is approached. The signature of the normal-to-superconducting transition is a sharp drop in resistivity seen in both type-I and type-II superconductors; however, resistivity is not an equilibrium thermodynamic property, and it is not related to a derivative of the free energy. We, thus, cannot use these resistivity measurements to check for the existence or the order of the transition in the Ehrenfest classification. This requires the measurement of an equilibrium thermodynamic property, which in this case (where the transition can be caused by varying the magnetic field) is magnetization. For type-II superconductors, this changes smoothly at the superconducting transition and, despite the sudden sharp drop in resistivity the transition is second order.

As was shown by Abrikosov, these infinitely long vortices in type-II superconductors form a two-dimensional hexagonal lattice in the superconducting state well below the transition temperature. A unique feature of this lattice is that the vortices are repelling each other—this solid is formed under the force of repulsion! Until the discovery of high-T_C superconductors, not much thought was given to possible transitions in this two-dimensional solid, assuming that it forms immediately below T_C. In the high-T_C superconductors, the operating temperature is high and thermal fluctuations were expected to play an important role at temperatures below the superconducting T_C. The importance of such fluctuations is enhanced by the small coherence volume of the paired electrons since the number of Cooper pairs in this coherence volume is accordingly small. This raised the interesting possibility of the vortex lattice melting through a first order phase transition, but the normal to superconducting transition is second order. Theoretical works predicted that the vortex melting transition could also be second order. Experimental observation of the vortex lattice melting and of determining the order of this phase transition became an important topic for researchers, and it was eventually established as a first order transition.

The Clausius–Clapeyron equation for a magnetic first order transition has been obtained following Equations 1.7 and 1.8. On the phase transition line

(H_C, T_C), the discontinuous jump in magnetization ΔM is related to the latent heat and the slope of the phase transition line by

$$dH/dT = -L/(T\Delta M). \tag{1.11}$$

This is applicable for all first order transitions caused by a varying magnetic field, including for vortex lattice melting. We describe briefly in the following text the experimental difficulties encountered in establishing relation (1.11) for this vortex melting transition. The experimental effort involved in establishing the nature of the vortex melting transition straddled the Ehrenfest and the modern classifications, and this is discussed in great detail in Chapter 3.

The vortex lattice to vortex liquid transition below T_C in BSCCO-2212 was found, by Zeldov et al. [10], to be accompanied by a discontinuous rise in magnetization. They scanned this transition with both a rising temperature and a rising magnetic field. The temperature was scanned for about 30 values of a fixed magnetic field, and the field was scanned for a similar number of fixed values of temperature. The isothermal field scans were performed at temperatures varying from about 90 K to about 40 K. (The transition at lower temperatures is from a vortex liquid to a disordered vortex solid through a second order phase transition, with the critical point near 40 K separating the first and second order transitions.) We should mention that the isothermal variation of pressure at such temperatures is experimentally quite difficult, but the isothermal variation of magnetic field presents no experimental difficulty. Both the isothermal scans of H with T between 40 and 90 K and the isofield scans of T with H below about 400 Gauss showed discontinuous jumps on the same phase transition line, providing a consistency test. The discussion persisted in view of theoretical works predicting that the vortex lattice melting transition should be second order, and the issue was settled only when Schilling et al. [11] measured the latent heat across the vortex lattice melting transition in $YBa_2Cu_3O_7$ (YBCO). They measured the magnetization jump in isothermal scans of magnetic field at various temperatures in the range 80–90 K. The transition was at higher fields (about 9 Tesla at the lowest temperature of 80 K) in contrast to earlier studies on BSCCO (below 0.04 Tesla). The latent heat $L(T)$ was estimated from the jump $\Delta M(T)$ seen at $H_m(T)$ in isothermal scans at various T as

$$L(T) = -T\Delta M(T)\, dH_m(T)/dT. \tag{1.12}$$

Direct measurements of latent heat were done by Schilling et al. [11] at about 15 temperatures by isothermal scans of H. This, again, involved scanning H to about 9 Tesla at temperatures around 78 K. As a third estimate of L, latent heat was directly measured by scanning temperature at a few values of magnetic field. All three estimates of latent heat agreed within experimental error. So much effort was needed to establish that the vortex melting transition was indeed a first order transition!

1.4 Studying Phase Transitions with Two Control Variables

In the Introduction, we discussed phase transitions caused by a variation in temperature. Natural phenomena are observed under ambient pressure, but we did discuss common experiences of humans that occur as pressure varies slightly from ambient. Similarly, magnetic field is not a variable in phenomena that we observe in nature, but there are phase transitions observed under even small variations in magnetic field. Physicists use both pressure and magnetic field as common control variables for, among other things, causing phase transitions. We discussed in the last section phase transitions observed by varying the magnetic field to test or establish new concepts in physics. Specifically, we used two variables to test the validity of the Clausius–Clapeyron equation and establish that the phase transitions were actually first order.

Phase transitions occur across a line in the space of two control variables, usually P and T. Magnetic transitions occur in the space of H and T, and some of these are also caused by pressure changes. We usually discuss the evolution of the state, and of G, under the variation of only one control variable, but the evolution of a state as one traverses arbitrary paths in two variables' space has been very important in understanding whether the system is in an equilibrium state. One early example of this was the discovery of the Meissner effect in superconductors.

The schematic in Figure 1.4 brings out that the superconducting state transforms to the normal state on heating in constant H and also on raising the field isothermally. The phase transition line separates a region of infinite electrical conductivity or zero resistance from a region of metallic (or higher) resistance. In the region of zero resistance, there is no electric field. It follows from Maxwell's equation that $dB/dt = 0$, i.e., the magnetic flux contained in the perfect conductor cannot change with time. Figure 1.4c shows four paths of reaching a particular value of (H_0, T_0) where the sample is superconducting. In path-1 the sample is in zero applied field when infinite conductivity sets in, and any subsequent increase in the applied H will keep $B = 0$. In path-2 the sample acquires infinite conductivity when $B = H_0$, and this value of B will be retained as long as it remains in the superconducting phase. In path-3 the sample acquires infinite conductivity when $B = H_C(T_0)$, and this value of B will be retained as long as it remains in the superconducting phase. In path-4 the sample acquires infinite conductivity at $T = 0$ when $B = H_C(0)$, and this value of B will now be retained as long as it remains in the superconducting phase. It is obvious that by choosing a suitable temperature T for entering the superconducting phase, the trapped flux inside the superconductor is fixed at $H_C(T)$ and thus the trapped flux at (H_0, T_0) can be tuned to any value between 0 and $H_C(T = 0)$. This path-dependence of B, and thus of the state at a particular (H, T) point, challenged the idea of the superconducting phase being a thermodynamic equilibrium state. The discovery of the Meissner effect, about two decades after the discovery of superconductivity, first gave a path-independent value to B. Meissner and Oschenfeld had shown that magnetic flux is totally expelled from the bulk of a sample when it makes the transition from the normal (N) to the superconducting (S) state, and $B = 0$, irrespective of the path followed to reach (H_0, T_0).

Let us assume that the material is a paramagnet in the normal state, with $\chi = 0$. The applied field fully penetrates into the sample, with $B = H_{app}$. As soon as T drops to T_C, the flux penetrating the sample drops to zero in accordance with the Meissner effect. The magnetization changes suddenly at T_C, from 0 to $-H_{app}$. The discontinuity in magnetization is, thus, zero on cooling in $H_{app} = 0$ and increases in magnitude as H_{app} is raised. Since the N → S transition is suppressed at $H_{app} = H_C(0)$, the discontinuity again vanishes. The N → S transition is thus second (or higher) order for $H_{app} = 0$ but is first order for all finite values of H_{app} below $H_C(0)$. The discontinuous drop in magnetization makes **B** = 0 in the entire (H, T) region where the superconducting phase exists, making it path-independent. The use of H as a second control variable brought out that the onset of superconductivity cannot be a phase transition if the only physical change is the onset of infinite conductivity; the Meissner effect allowed this N → S transition to be accepted as a thermodynamic transition between two equilibrium phases. We now consider another drastic result obtained by the use of two control variables for a first order phase transition.

In Figure 1.3c we showed a schematic of the melting curve of helium-3, another unusual first order transition. The use of pressure as the second control variable in this transition results in a method of cooling called Pomeranchuk cooling, which is used at very low temperatures. We first understand that phase coexistence on a first order phase transition line allows temperature in a confined volume to vary just by varying the second control variable. We shall briefly describe how this use of pressure—to cool and lower the temperature in experiments planned to study the solid phase of helium-3—resulted in the unintended discovery of a new phenomenon in the liquid phase of helium-3, with a subsequent Nobel Prize.

To understand Pomeranchuk cooling, we take a cell containing only liquid helium-3, isolate this cell thermally at a temperature below 319 mK, and start raising the pressure. At the appropriate pressure, the liquid will start freezing, the isolated system cannot maintain a fixed temperature, and the liquid and solid can coexist only on the melting line. As the pressure is raised further, more liquid solidifies but, because of the latent heat and consequent phase coexistence associated with the first order transition, the mixture is constrained to remain on the melting line as the pressure is increased. This is true for any liquid that is less dense than solid, and the slope of the freezing line is positive in all such cases with the exception of helium-3 below 319 mK. In all cases where the solid has a higher density than the liquid, except for helium-3, the temperature of the liquid–solid mixture will rise as pressure is raised and the mixture will eventually become entirely solid. For helium-3, however, the liquid has lower entropy and latent heat is absorbed as the liquid freezes, and the temperature of the liquid–solid mixture will drop even as the mixture eventually becomes entirely a solid. (For the case where the liquid has higher density, the solid melts on increasing pressure. If we have a complete solid or a liquid–solid mixture in such cases, then the solid will melt and the temperature of the mixture will fall. Here, however, the mixture will eventually become entirely a liquid.) So if we start with a mixture of solid and liquid helium-3 below 319 mK temperature, then it is on the melting curve in the anomalous region where the solid melts on cooling. The increasing pressure transforms the liquid into solid, with the temperature of the mixture dropping even though the liquid does have higher volume. This process of cooling is called Pomeranchuk cooling, after the physicist who theoretically proposed its occurrence. This cooling, resulting from increasing pressure, will continue until all the liquid has transformed to solid, at which temperature the coexistence will end and we will have only solid in the cell, and then pressure will rise isothermally. This provides a method of cooling from 300 mK down to below 3 mK, and the cooling can continue to the lowest temperature by starting with liquid at low enough temperature. Osheroff, Richardson, and Lee were studying low temperature properties of solid helium-3. They were using Pomeranchuk cooling, and even at about 2 mK the sample cell contained both solid and liquid phases of helium-3. They observed interesting kinks in the time evolution of the pressure in the Pomeranchuk cell, which they first attributed to a possible magnetic transition in solid helium-3, in a paper titled "Evidence for a new phase of solid He^3" [12]. They soon realized that these kinks could actually be originating from the coexisting (parasitic) liquid phase. Within a period of about five months, they established that the origin of the kinks was actually a phase transition in the liquid phase that was coexisting in the cell undergoing Pomeranchuk cooling [13]. For this, they used a one-dimensional version of what is now called MRI. This phase transition was shown to be the onset of superfluidity in liquid helium-3, and they received the Nobel Prize in 1996. The onset of superfluidity in liquid helium-3 was thus discovered fortuitously during a study on solid helium-3, while following the phase coexistence at a first order transition with pressure as the second control variable.

The two examples discussed bring out that a phase transition should not be considered only to occur when temperature is varied: it must also be observed under variation of another control variable. Studying it with all the control variables that can cause phase transition is not only necessary for testing compliance with the Clausius–Clapeyron relation but the results so obtained provide better understanding of the physics. One general condition is that the transition temperature T_C does vary with variations in the second control variable. We can also exploit the second control variable when there is another characteristic temperature in the problem (say T_k), whose dependence on the second control variable is different from that of T_C. The physics underlying the two characteristic temperatures would be different and will manifest more prominently experimentally when we can reduce or enhance the difference between these two temperatures. The different dependences of T_C and T_k on the second control variable allow us to do this. We shall return to the importance of this in subsequent chapters.

We emphasize that varying both the control variables is one novelty of the studies that form the essence of this book. Another new feature appears when the transition temperatures depend on two more control variables, like pressure and magnetic field. An example of this was the study by Kushwaha et al. [14] on Pd-doped FeRh where the first order magnetic transition, between the FM and the AFM phases, was made to occur at the same transition temperature T_C for different combinations of (P, H).

While phase transitions are usually discussed with temperature as the control variable, we now discuss an exactly solvable model of a first order phase transition, namely, the van der Waals gas model, which was proposed in the late nineteenth century to explain the behavior of real gases and for which van der Waals received the Nobel Prize in 1910. This model is usually discussed as solved on an isotherm, with pressure as the control variable. In keeping with the earlier discussion on the importance of studies with two control variables, we shall discuss its solution on both isotherms and isobars. This model is an exactly solvable model describing the first order transition between a liquid and gas. We will discuss it in detail since it also allows us to understand limits for the existence of metastable states associated with this first order transition.

1.5 Van der Waals Gas as an Exactly Solvable Model: Isotherms and Isobars

This model is more than a century old and was an initial step to go beyond the ideal gas model and include interactions between particles. The gas particles were assigned a finite volume that was subtracted from the total volume available in the ideal gas law. Additionally, an attractive force between the gas particles was included by modifying the pressure term in the ideal gas law so that the effective pressure was higher for a higher density of particles. This simple model gave a better approximation to the behavior of real gases than did the ideal gas law, and its success was in describing the behavior of a gas as the temperature or pressure was varied. This model could also predict the liquefaction of a gas and the metastable states around this liquid–gas transition. We shall concentrate on this additional feature of the model.

The van der Waals gas is an exactly solvable model of this first order phase transition, though the exact solution may be obtained only numerically. We shall use this model to identify questions in the context of first order transitions that remain unaddressed in the Ehrenfest classification, specifically issues related to metastable states across a phase transition. This model predicts the behavior of a gas in the supercooled state and that of the liquid in the superheated state, and it further provides limits after which these metastable states become unstable. We will study this model in detail in view of our interest in metastable states associated with first order transitions.

The van der Waals equation is

$$\left(P + a/V^2\right)\left(V - b\right) = RT \tag{1.13}$$

On an isotherm (i.e., for fixed T), this is a cubic equation in V that can be solved for V(P). This is the commonly discussed form of the van der Waals equation. The equation can also be solved on an isobar (fixed P) for V(T), though this is not the commonly discussed form. One reason for this is that P and V are conjugate, in that V is the derivative of G with respect to P, and we shall see that their product having units of energy paved the basis for calculating the transition temperature following Maxwell's construction. Temperature is, however, more commonly used as a control variable and has importance for our discussion on metastable states. We shall solve the van der Waals equation with both the situations—of a fixed T with P as the control variable and a fixed P with T as the control variable. This is in continuation of our assertion that first order phase transitions should be discussed in the context of at least two control variables.

The cubic equation (1.13) for V as a function of P can be solved numerically for given values of the parameters "a" and "b." Since all coefficients are real, the equation can have either three real roots or one real and two complex conjugate roots. Phase coexistence at a first order phase transition would correspond to the existence of two physically valid real roots.

The solutions for V(P) at various constant values of T in this intermediate region are as depicted schematically in Figure 1.7a. The high-P and low-V region corresponds to the behavior of the liquid, while the low-P and high-V region corresponds to the gas. At pressures above a certain critical value, there is only one real solution, and this corresponds to a liquid. This critical value (P*) depends on the constant temperature of the isotherm; as pressure is lowered below this P* on the isotherm, the liquid reaches the interim region where there are three real solutions to the van der Waals equation and one sees a kink in the V(P) curve. There are now three solutions for each value of P—corresponding to the liquid volume V_L, an intermediate volume V_X, and a gas volume V_G—and any linear superposition of these three is also a mathematical solution. The "physical" solution is a linear superposition of only V_L and V_G because V_X would have negative compressibility in that the volume is directly proportional to pressure at that point V_X. Volume increasing with increase in pressure is not physically acceptable, and the solution V_X is not a physical solution.

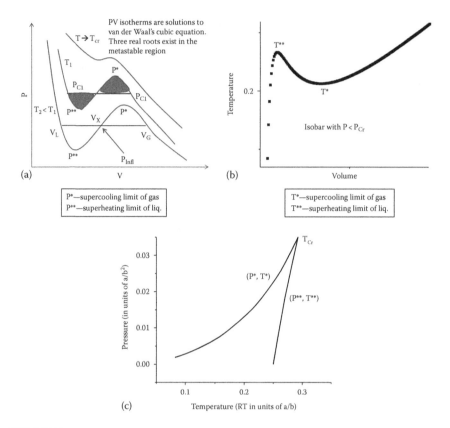

(a)

(b)

P*—supercooling limit of gas
P**—superheating limit of liq.

T*—supercooling limit of gas
T**—superheating limit of liq.

(c)

FIGURE 1.7
Solutions to van der Waals equation are depicted. In (a) we show isotherms, where P-V curves are shown for fixed temperature, for a few typical temperatures. P_{C1} corresponds to the phase transition pressure for the isotherm at T_1 as per Maxwell's construction [15]. The minimum and maximum are at (P**, V**) and at (P*, V*), respectively, and the inflection point is at (P_{infl}, V_{infl}). In (b) we show isobars, where T-V curves are shown for fixed pressure, for one typical pressure. The minimum and maximum are at (T**, V**) and at (T*, V*), respectively, and the inflection point is always at V = 3b. In (c) we show the plot of (T**, P**) and (T*, P*), obtained by choosing various values of V and then using Equations 1.15a and 1.15b as discussed in the text. Note that the limits for supercooling and superheating are obtained analytically, whereas the transition point cannot be obtained analytically.

The first order liquid-to-gas phase transition occurs when the solution corresponding to only liquid (V_L) transforms into a solution corresponding to superposition of liquid and gas [$cV_L + (1 - c)V_G$]. While this transformation can take place at a pressure just below P* when multiple real solutions start existing, it does not become energetically favorable until it reaches a pressure P_C corresponding to the Maxwell construction [15]. At P_C, the horizontal constant-P line encloses an area on the left limited below by the minimum in V(P), which is equal to the area on the right enclosed by the horizontal line and limited above by the maximum in the V(P) curve. At this pressure

$P_C(T)$, the liquid with volume V_L on the left curve starts converting to gas with larger volume V_G on the right curve. There is a sudden increase in the volume, corresponding to the first order transition, and gas and liquid coexist with the pressure remaining constant as the liquid fraction c reduces from 1 to 0 and the entire liquid converts to gas. The pressure then reduces further and the volume of the gas increases on the isotherm. This is the well-known equilibrium solution corresponding to the minimum in energy. As the temperature of the isotherm is raised, the discontinuous increase in the volume reduces, until at a critical temperature T_{Cr} the change from liquid to gas is continuous with no volume jump or latent heat—there is no first order phase transition. In the following, we shall solve the van der Waals equation to obtain T_{Cr}.

For all isotherms that have three real roots, we can cause liquid–gas transition by varying pressure, and for all such isotherms, there is a pressure P^* above which only the liquid phase exists. These isotherms also have a critical pressure P^{**} below which only the gas phase can exist. As we shall see in the following text, P^{**} can be negative for isotherms corresponding to low temperatures, and it then loses physical significance. There is another mathematically significant pressure P_{infl} at which V(P) has an inflection point. We shall discuss in the following text whether this pressure has any physical significance.

Let us follow the solution to the cubic equation on this isotherm as we lower the pressure. Suppose that the gas is not formed at P_C when the liquid volume is V_L. The volume of the liquid keeps increasing as pressure reduces below P_C, and we still have a physically acceptable solution corresponding to a liquid whose volume is increasing with reducing P. When the liquid volume has increased to V^{**} as P reaches the minimum value P^{**}, we reach the unphysical situation that any further smooth increase in volume on the V(P) curve corresponds to a material with negative compressibility whose volume is increasing with an increasing P. At this stage, the only real solution with $P < P^{**}$ corresponds to the gas, and there must be a transition with a sudden sharp increase in volume. The liquid cannot exist at $P < P^{**}$ or with volume $V > V^{**}$.

The liquid that exists for $P^{**} < P < P_C$ is not in equilibrium because it is not the minimum energy configuration: it is a metastable state and is a "super-heated liquid." This minimum at P^{**} has an important significance in the following discussion as the limit of metastability at which the superheated liquid must convert to the gas. We shall also solve the van der Waals equation to obtain P^{**} and then recognize that P^{**} reduces as T is reduced, and finally becomes negative and unphysical.

Let us now follow this isotherm from the large V and low pressure end, where there is only one real solution corresponding to gas. The volume of the gas decreases as pressure is raised. As the pressure rises above P^{**}, we have three solutions, corresponding to the liquid volume V_L, an intermediate volume V_X, and a gas volume V_G. The first order gas-to-liquid phase transition occurs when the pure gas solution transforms into a superposition of liquid and gas. While this transformation can take place at any pressure above P^{**} (at a low T where P^{**} is negative, this can happen at any P), it does not become energetically favorable until P rises to the P_C corresponding to the Maxwell construction. At this pressure $P_C(T)$, the gas with the larger volume V_G on the right curve starts converting to the liquid with volume V_L on the left curve. There is a sudden decrease in volume, corresponding to first order transition, and the gas and liquid coexist with the pressure remaining constant as the entire gas converts to liquid. This is a solution of the van der Waals equation corresponding to superposition of liquid and gas, i.e., $[cV_L + (1 - c)V_G]$, where the gas fraction $(1 - c)$ reduces from 1 to 0 before P starts rising again. Assuming that the liquid does not form at P_C, we still have a solution with $c = 0$ corresponding to a gas whose volume is decreasing as the pressure is raised above P_C. As V reduces and P reaches the maximum value P^*, we reach the unphysical situation of the V(P) curve corresponding to a material with negative compressibility. At this stage, the only real solution with $P > P^*$ corresponds to liquid, and there must be a transition with a sudden change in volume. The gas that exists for $P^* > P > P_C$ is not in equilibrium because it is not the minimum energy configuration: it is a metastable state and is a "supercooled gas." This maximum at P^* is the limit of metastability at which the supercooled gas must convert to liquid. We shall also solve the van der Waals equation to obtain P^*. Somewhere between P^* and P^{**} on the unphysical negative compressibility part of the isotherm is a point P_{inf} where V(P) has an inflection point.

In the region where the van der Waals equation has three real roots, there is one maximum $P^*(T)$, one minimum $P^{**}(T)$, and one inflection point $P_{inf}(T)$. Let us solve Equation 1.13 to obtain these important points in P as a function of V. The van der Waals equation can be solved completely only numerically, but we can analytically obtain P^*, P^{**}, and P_{inf} using the following protocol. We first choose a value of V, and then solve for T such that this value of V corresponds to P^*, or P^{**}, or the inflection point at that T.

Using Equation 1.13, we note that

$$P + a/V^2 = RT/(V - b)$$

and then

$$dP/dV = 2a/V^3 - RT/(V - b)^2 \tag{1.14}$$

This first derivative $dP/dV = 0$ when P(V) has a minimum (as at P^{**}) or a maximum (as at P^*), and the limit of metastability is for an isotherm at T that satisfies

$$RT = 2a(V - b)^2/V^3 \tag{1.15a}$$

Thus, for any choice of V, Equation 1.15a gives a temperature at which this V corresponds to the limits of metastability either of the liquid (V**) or of the gas (V*). The corresponding values of pressure would be now obtained by solving (1.13):

$$P = a(V - 2b)/V^3 \tag{1.15b}$$

This is the pressure at which one obtains a maximum or a minimum in V(P) at that V with T given by (1.15a). Whether this combination of V and T has given a P that corresponds to P* or P** can be confirmed by the sign of

$$d^2P/dV^2 = -6a/V^4 + 2RT/(V - b)^3 \tag{1.15c}$$

being positive at P** and negative at P*. After slight manipulation of (1.15c) under the constraint of (1.15a) for the value of the second derivative at a point where dP/dV vanishes, we get the following:

$$d^2P/dV^2 = -6a/V^4 + 2RT/(V - b)^3 = \left[4aV - 6a(V - b)\right]/\left[V^4(V - b)\right]$$
$$= \left[2a(3b - V)\right]/\left[V^4(V - b)\right] \tag{1.15d}$$

This is positive for V < 3b and (1.15b) would give a minimum corresponding to P**, and this is negative for V > 3b so that (1.15b) would give a maximum corresponding to P*, in the P-V isotherm. *We thus choose a value of V, use (1.15a) to obtain the isotherm T at which this corresponds to an extremum in V(P), use (1.15b) to obtain the pressure corresponding to that extremum, and from (1.15d) decide that it is P** for V < 3b and P* for V > 3b.* We now obtain the inflection point, where we must have

$$d^2P/dV^2 = 0 \tag{1.15e}$$

This must be without the constraint of (1.15a) or, consequently, of (1.15b), which gives that V is an inflection point at

$$RT = 3a(V - b)^3/V^4 \tag{1.15f}$$

We emphasize that the inflection point is at different V for different isotherms. Specifically, the inflection point is at V = 2.5b for RT = 162a/(625b) and rises monotonically to V = 3b at RT = RT_C = 8a/(27b).

Both (1.15b) and (1.15f) have to be satisfied at the critical point where the minimum and the maximum merge with the inflection point. So

$$a(V - 2b)/V^3 = \left[3a(V - b)^2 - aV^2\right]/V^4$$

or

$$V(V-2b)=3(V-b)^2-V^2 \quad \text{or} \quad V^2-4bV+3b^2=(V-3b)(V-b)=0$$

Since $V = b$ corresponds to a negative P, the maximum, minimum, and inflection point merge at

$$V_{Cr} = 3b \tag{1.16}$$

Substituting this back in (1.15b), or in (1.15f), we get the critical pressure:

$$P_{Cr} = a/27b^2 \tag{1.17}$$

Substituting in Equation 1.15a, we get the critical temperature:

$$RT_{Cr} = 8a/27b \tag{1.18}$$

Beyond this critical point, i.e., at $T > T_{Cr}$, there is no first order transition and the coexistence of liquid and gas phases is not observed in the van der Waals model. Above this critical temperature, there is only one real solution for all values of P.

We have identified P*, to which pressure can be raised without liquefying the gas, as the limit of metastability (or of supercooling) of the gas and P**, to which pressure can be lowered without converting the liquid to gas, as the limit of metastability (or of superheating) of the liquid. We can calculate these on various isotherms with $T < T_{Cr}$. The procedure is totally analytical and exact, independent of the values of "a" and "b." As examples, we list in Table 1.1 the results of P* (or P**) and the corresponding T for a few values of V.

We thus obtain the metastability limits (of supercooling and superheating) analytically, and these are depicted in Figure 1.7c. We must emphasize here that unlike P*(T) and P**(T), the phase transition line $P_C(T)$ cannot be obtained analytically. This figure is a phase diagram for the van der Waals equation, showing limits to which a liquid can exist before converting to a gas and the limits to which a gas can exist before converting to a liquid. These lines are solutions to $dP/dV = 0$ for various T.

We stress here that except for the critical point, the solutions to $d^2P/dV^2 = 0$ are not shown. We shall argue, by comparing the solutions of the equation on isotherms and on isobars, that these inflection points have no physical significance.

We also note by substituting Equation 1.15a in (1.13) that

$$P = RT/(V-b)-a/V^2 = 2a(V-b)/V^3-a/V^2 = a(V-2b)/V^3$$

TABLE 1.1

The Limits of Metastability as Obtained for Each V Using Equation 1.15a
or 1.22b, and Using Equation 1.15b or 1.22a

V (Units of b)	RT (Units of a/b)	P*/P** (Units of a/b²)
V	$RT = 2a(V-b)^2/V^3$	$P = a(V-2b)/V^3$
1.25	12/625	−192/625
1.5	4/27	−4/27
2	1/4	0
7/3	96/343	9/343
2.5	36/125	4/125
8/3	75/256	9/256
3	**8/27**	**1/27**
4	9/32	1/32
5	32/125	3/125
6	25/108	1/54
7	72/343	5/343
10	152/1,000	8/1,000
20	361/4,000	9/4,000
100	1782/1,000,000	8/1,000,000

For V < 3b these correspond to the limit (T**, P**) to which the liquid can be superheated, and for V > 3b these correspond to the limit (T*, P*) to which the gas can be
supercooled. The values in bold for V = 3b correspond to the critical point T_{Cr}.

Thus, P(V) goes negative for isotherms where the minima occur at V < 2b.
These provide examples of solutions for a fixed isotherm that are real but
may be unphysical in that the value of P is negative. Such real but unphysical solutions are not encountered when solving the van der Waals equation
on isobars.

These results were obtained by varying P isothermally. An analysis of
P(V) on isotherms has helped provide justification for Maxwell's construction, but we have earlier stressed that phase transitions need to be studied
by varying both P and T as control variables. We now discuss *solution of
van der Waals equation on isobars* when T is varied with constant P. We can
solve the van der Waals equation to obtain V as a function of T for fixed
values of P.

$$RT = PV + (a/V) - (ab/V^2) - bP \tag{1.19}$$

$$R\, dT/dV = P - (a/V^2) + (2ab/V^3) \tag{1.20}$$

$$R\, dT^2/dV^2 = (2a/V^3) - (6ab/V^4) \tag{1.21}$$

And we note the simple conclusion that $dT^2/dV^2 = 0$ for $V = 3b$, and *the inflection point in T(V) on isobars is fixed at V = 3b for all values of P. This contrasts with the result (1.15f), where the location of the inflection point on isotherms was dependent on T.* Specifically, the value of V corresponding to the inflection point on isotherms rises monotonically from $V = b$ as T is raised from 0, and $V = 3b$ is an inflection point only for the isotherm $T = 8a/(27b)$ corresponding to the critical point.

Since the inflection points occur at different values for isobars and isotherms, we reassert our statement that *these inflection points have no physical significance* even though they are obtained analytically. The entire range of solutions corresponding to negative compressibility (for isotherms) and to negative thermal expansion (for isobars) have no physical significance, and the inflection point is in this region of negative compressibility. We must check whether the limits of metastability occur at the same values of the control parameters for both isobars and isotherms, to support the view that these have a physical significance and correspond to thermodynamic limits.

We also note that the lowest value of P(V) for any isotherm was obtained when $dP/dV = 0$ and $V < 3b$, whereas the lowest value of T(V) for any isobar is obtained when $dT/dV = 0$ and again $V < 3b$. Equation 1.20, however, gives $dT/dV = 0$ at $P = a(V - 2b)/V^3$, which is the same as Equation 1.15b obtained from considering isotherms. Since P is nonnegative on the isobars, we do not access the region $V < 2b$ where we had $P^{**} < 0$ and that was accessed while discussing the conventional solution of the van der Waals equation with isotherms. We thus persist with this discussion of the van der Waals equation with isobars, i.e., with T (rather than P) as the control variable.

We will get solutions for T(V) as depicted in Figure 1.7b, and all the earlier discussion using isotherms can be followed through. At large T and large P, we have only one real solution, but at $P < P_{Cr}$ we have three real solutions for all $T < T_{Cr}$. The solution with large V corresponds to gas, the one with small V corresponds to liquid, and the one with intermediate V is unphysical as it has a negative thermal expansion. We must mention here that such a negative thermal expansion is observed in some solids, but the van der Waals model has no provision for this behavior and thus the region of negative thermal expansion has to be discarded as unphysical.

Let us follow the solution to the cubic equation closely on the isobars. As we start raising temperature from the low-T and low-V point on an isobar with $P < P_{Cr}$, the temperature rises and the volume of the liquid increases on the isobar. The transition temperature T_C would be determined by an analog of the Maxwell construction, but there is no obvious analogy with the isotherm treatment because the area enclosed by a P-V curve does correspond to energy whereas the area enclosed by a T-V isobar does not even have the units of energy. Nevertheless, the transition temperature does lie between T^* and T^{**} and the liquid can exist in a metastable state above T_C to the maximum in the isobar at T^{**}. This region represents a superheated liquid.

Let us now follow this isobar from the large-V and high-T end, where there is only one real solution corresponding to gas. The volume of the gas decreases as T is lowered. As the temperature drops to T**, we now have three solutions, corresponding to the liquid volume V_L, an intermediate volume V_X, and a gas volume V_G. The first order gas-to-liquid phase transition occurs when the pure gas solution transforms into a superposition of liquid and gas. While this transformation can take place at any T below T**, it does not become energetically favorable until it reaches the T_C corresponding to some counterpart of the Maxwell construction. Assuming that the liquid does not form and the temperature is lowered below T_C, we still have a solution corresponding to a gas whose volume is decreasing, but as T reaches the minimum value T*, we reach the unphysical situation of the V(T) curve corresponding to a material with negative thermal expansion. This minimum also corresponds to the limit to which the gas can be supercooled.

We have the liquid transforming to gas with a sudden increase in volume with rising T, the gas and liquid coexisting until the entire liquid converts to gas. This is the equilibrium phase transition, but the van der Waals model does not estimate the latent heat. The model does tell that the liquid can be superheated above T_C to the maximum at T** and the gas can be supercooled below T_C to the minimum at T*. These correspond to maximum and minimum in T(V) and should satisfy dT/dV = 0, and we get from Equation 1.20

$$P - a/V^2 + 2ab/V^3 = 0 \qquad (1.22a)$$

$$RT = 2a(V - b)^2/V^3 \qquad (1.22b)$$

We note that (1.22a) is identical to (1.15b) and (1.22b) is identical to (1.15a). Since V = 3b is an inflection point for T(V) for all P, we note that dT/dV = 0 corresponds to the maximum in T(V) for V < 3b and represents the limit of metastability of the liquid and to a minimum for V > 3b representing the limit of metastability of the gas. Both isotherms and isobars yield identical limits of metastability though the inflection points given by Equations 1.15f and 1.21 are not identical. On isotherms, (1.15f) reduces to (1.16) only when (1.15a) is also satisfied, when the inflection point coincides with an extremum.

Solutions of the van der Waals equation on isobars and isotherms have provided a simple analytical prescription to obtain the limits of supercooling and superheating. Both give identical results for the limits of metastability, clearly independent of the path followed in the control variable space. This reinforces the view that the limits of metastability, corresponding to superheating of the liquid and supercooling of the gas, are physically meaningful limits.

Solutions on isobars also allow us to calculate the critical point where the minimum and maximum merge. This satisfies both dT/dV = 0 and

$d^2T/dV^2 = 0$. From Equation 1.21, the latter gives us $V_{Cr} = 3b$, as in Equation 1.16. Substituting in Equation 1.22a, we get $P_{Cr} = a/(27b^2)$, as in Equation 1.18. Substituting in Equation 1.22b, we again get $RT_{Cr} = 8a/27b$. The consistency between using P as the control variable and using T as the control variable is thus complete for this exactly solvable model of a first order phase transition. Specifically, this is also true for the limits of the metastable regime.

1.6 Metastable States across Phase Transition: Limitations of the Ehrenfest Classification

The behavior depicted by the van der Waals model between T_C and T^* during cooling—and between T_C and T^{**} on heating—corresponds to states that are not equilibrium states (Figure 1.8). They have a free energy that is higher than that of the equilibrium state (liquid and gas, respectively, in the two temperature regimes) and are accordingly not stable states. At this stage, we cannot explicitly show whether they are stable against small perturbations and could be called metastable or not stable under even infinitesimal perturbations and should be called unstable. We do not have a prescription to test for these two possibilities. If they are metastable, then is there a lower limit for the magnitude of perturbation that makes them unstable? Is this limit dependent on how far the control variable T is from the critical value T_C for the phase transition? Is the existence of such metastable states generic to first order transitions?

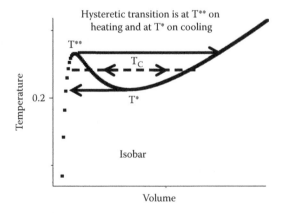

FIGURE 1.8

The van der Waals model predicts an equilibrium liquid–gas transition at T_C. It also predicts the possibility of a hysteretic transition with liquid to gas occurring at T^{**} and the reverse gas to liquid occurring at T^*.

We refer to Figure 1.5 to discuss a generic phase transition. The inequality between the free energies of the two phases, viz., G_1 and G_2, changes sign at T_C. Supercooling corresponds to the state persisting in phase-2 with $G = G_2(T)$ even for $T < T_C$, without transforming to the equilibrium phase with lower free energy at that T. Similarly, superheating corresponds to the state persisting in phase-1 with $G = G_1(T)$ even for $T > T_C$, not transforming to the equilibrium phase with lower free energy $G_2(T)$ at that T. The existence of a phase transition requires that $G(T)$ is not analytic at T_C, irrespective of the order of the phase transition. Thus, analytic continuation of $G_1(T)$ into the regime $T > T_C$ will result in a value higher than $G_2(T)$ irrespective of the order of the phase transition, as depicted in Figure 1.6. We thus have the following additional open issues or unresolved problems with the Ehrenfest scheme.

Problem 1: Supercooling and superheating do not appear to be disallowed for a phase transition of arbitrary order. Folklore, based on observations, is that it can take place for only first order phase transitions. This needs to be settled emphatically.

In our discussion on the van der Waals equation, we noted that supercooling and superheating can only persist till a particular value of pressure (on isotherms) or temperature (on isobars) away from the phase transition point. The generic discussion of the state continuing on the $G(T)$ corresponding to the nonequilibrium phase, depicted in Figure 1.5, provides no reason for supercooling or superheating to be terminated for a particular excursion beyond the phase transition point [7]. This supercooling to an arbitrary extent is contrary to some established phenomenological facts on the supercooling of water. While it is difficult to supercool water that is being stirred, undisturbed water may be supercooled a few degrees. Supercooling of pure water in a container with very smooth internal surface is possible to a lowest temperature of about $-40°C$. This can be formulated as the second unresolved problem.

Problem 2: There must be a generic explanation for a limit to supercooling, and to superheating.

Another phenomenological observation is that when supercooled water is disturbed, it freezes. The disturbance required is more when the extent of supercooling is less: a deeply supercooled liquid freezes under a smaller disturbance. There is no indication of this behavior even from the van der Waals equation. Since supercooled and superheated states are metastable, they must relax with time to the respective stable states. A corollary to the phenomenological observation is that a more deeply supercooled state, which is at a lower temperature, must have a higher relaxation rate and relax faster to the equilibrium state.

Problem 3: There is no prescription for quantifying a "level of metastability" or estimating the relative relaxation rates of different supercooled states, or of different superheated states.

In Section 1.3, we discussed in Equation 1.10 the case when a derivative of free energy is continuous but diverges logarithmically. We noted this as a major drawback of the Ehrenfest classification, which led to the modern classification. In this section, we have discussed some observed behavior of metastable states that cannot be addressed within the Ehrenfest classification. As noted in the book *Statistical Mechanics* by Kerson Huang, "Thermodynamics is a phenomenological theory of matter. As such, it draws its concepts directly from experiments" [16]. We thus recognize the need for a different perspective on phase transitions, stressing the need to understand metastable states. A perspective has existed for some time that overcomes the need to classify transitions satisfying Equation 1.10 and will be described in Chapter 2. We shall bring out how this answers the three issues raised as "Problems" in this section. This will enable us to look at more recent experimental results on materials that do not enable measurement of latent heat and where disorder effects introduce new observations.

References

1. V.S. Ramachandran, *The Tell-Tale Brain: A Neuroscientist's Quest for What Makes Us Human*, W.W. Norton & Company, New York (2011).
2. A. Steyn-Ross and M. Steyn-Ross (Eds.), *Modeling Phase Transitions in the Brain*, Springer, New York (2010).
3. https://en.wikipedia.org/wiki/Heat.
4. W. Meissner and R. Ochsenfeld, *Naturwissenschaften* **21** (1933) 787.
5. H.K. Onnes, *Commun Phys Lab Univ Leiden* **12** (1911) 120.
6. P. Kumar, D. Hall, and R.G. Goodrich, *Phys Rev Lett* **82** (1999) 4532.
7. K. Huang, *Statistical Mechanics*, Wiley Eastern, New Delhi, India (1975), p. 38.
8. R.M. White and T.H. Geballe, *Long Range Order in Solids*, Academic Press, New York (1979), p. 12.
9. C. Kittel, *Introduction to Solid State Physics*, Wiley Eastern, New Delhi, India (1971), pp. 352–353.
10. E. Zeldov, D. Majer, M. Konczykowski, V.B. Geshkenbein, V.M. Vinokur, and H. Shtrikman, *Nature* **375** (1995) 373.
11. A. Schilling, R.A. Fisher, N.E. Phillips, U. Welp, D. Dasgupta, W.K. Kwok, and G.W. Crabtree, *Nature* **382** (1996) 791.
12. D.D. Osheroff, R.C. Richardson, and D.M. Lee, *Phys Rev Lett* **28** (1972) 885.
13. D.D. Osheroff, W.J. Gully, R.C. Richardson, and D.M. Lee, *Phys Rev Lett* **29** (1972) 920.
14. P. Kushwaha, P. Bag, R. Rawat, and P. Chaddah, *J Phys: Condens Matter* **24** (2012) 096005.
15. K. Huang, *Statistical Mechanics*, Wiley Eastern, New Delhi, India (1975), pp. 42–45.
16. K. Huang, *Statistical Mechanics*, Wiley Eastern, New Delhi, India (1975), p. 3.

2

Modern Classification of Phase Transitions

2.1 First Order Transitions, and the Rest

The inability of the Ehrenfest classification to classify phase transitions where the first derivative of G is continuous at the transition point but the second derivative does not exist led to the modern classification, which we shall outline in this chapter. We have also shown in Chapter 1 that the Ehrenfest classification has three limitations related to supercooling and superheating. This aspect has not been given enough importance in earlier texts, mainly because of the comparative neglect of metastable states.

The modern classification, however, provides an understanding of metastable states. As we outline the modern classification, we shall emphasize metastable states and the resulting hysteresis for first order phase transitions, metastable states being the *raison d'être* of this book. We shall lead to how hysteresis, with suitable cross-checks, provides an experimental benchmark akin to latent heat and the Clausius–Clapeyron relation.

As discussed in Chapter 1, there will be situations where the nth derivative of free energy is the highest derivative that is continuous, and the (n + 1)th derivative cannot be described as discontinuous because it simply does not exist. Examples of this phenomenon are observed experimentally as in the λ-transition in liquid helium and the ferromagnetic transition at the Curie point, and the behavior of heat capacity in such a case was given by Equation 1.10. This aspect was highlighted by experimental measurements and has been given importance in earlier literature as the *raison d'être* for the modern classification based on the behavior of order parameters. The observed cases correspond to the derivative with n = 1 being continuous and n = 2 not being defined. The third and higher order transitions, as defined in the Ehrenfest scheme, have not been established experimentally. We thus have three categories of phase transitions that have been observed experimentally: (1) derivative of G with n = 1 is discontinuous, (2) derivative of G with n = 1 is continuous while derivative with n = 2 is discontinuous, and (3) derivative of G with $n = 1$ is continuous while derivative with $n = 2$ is not defined. Category (1) has latent heat associated with the transition, while categories (2) and (3) do not have latent heat accompanying the transition. Based on this, phase transitions are now categorized as those that have accompanying latent heat and those that do not. The former fall within the ambit of the Ehrenfest classification and are called first order phase transitions. The latter would cover "the rest" in an extension of the Ehrenfest classification and are now called continuous phase transitions. We elaborate on the implications of this emphasis on latent heat as the basis for classifying phase transitions.

The existence of latent heat implies that conversion from phase-1 to phase-2 will proceed over a finite time even after the control parameter (usually T) has reached its critical value (T_C). During this finite conversion time, the control parameter cannot change from this critical value and both phases will coexist. Phase coexistence is a necessary consequence of latent heat, and thus a characteristic of first order transitions.

In the absence of latent heat, the transition will proceed instantaneously on the material equilibrating at the transition point. Thus, the two phases do not coexist but one phase transforms instantaneously to the other. If we increase the temperature, then phase-2 forms on reaching T_C, and if we lower temperature, then phase-1 forms on reaching T_C. It follows rigorously that in the absence of latent heat, phase-1 and phase-2 must be identical at $T = T_C$ and that there is no coexistence of the two phases.

The coexistence of two phases is a signature of latent heat and, thus, of a first order phase transition. We thus have "phase coexistence" as another crucial feature that distinguishes first order phase transitions from continuous phase transitions.

Phase coexistence at the phase transition point implies that a first order phase transition must proceed by nucleation and growth of one phase in the matrix of the other. We characterize the two phases as having a physical parameter that is different, and which we call an order parameter. For a liquid-to-gas transition, this order parameter is the macroscopic density. For a first order phase transition, the two phases with different order parameters coexist at $T = T_C$, as depicted in Figure 2.1. It follows that the transition initiates at $T = T_C$ by nucleation of the second phase with a very different order parameter, i.e., with the onset of large change in order parameter but over small length scales. The correlation length is finite, and nuclei of the second phase are small. (We shall see later in the book that this small correlation length allows different regions to have different transition temperatures, permitting a first order transition to be broadened in real materials with quenched disorder.)

In a transition with no latent heat, the two phases become identical at $T = T_C$ and their order parameters merge smoothly at $T = T_C$, as depicted in Figure 2.2. The transformation is instantaneous at $T = T_C$ and must occur

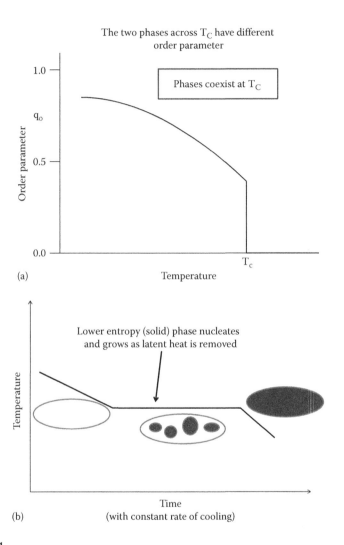

(a)

(b)

FIGURE 2.1
We show in schematic (a) the variation of order parameter q_0 as the control variable (temperature) is varied for a first order phase transition. There is a discontinuous change in q_0 at the transition temperature T_C. The two phases with distinct order parameters coexist at this point, and latent heat must be added (while heating) before the temperature can change. This is brought out in schematic (b), where we show the temperature variation under a constant cooling rate. The formation of the lower entropy (or higher order parameter) phase as the temperature is lowered proceeds by nucleation and growth during the time the temperature stays fixed at T_C, as brought out by the schematics.

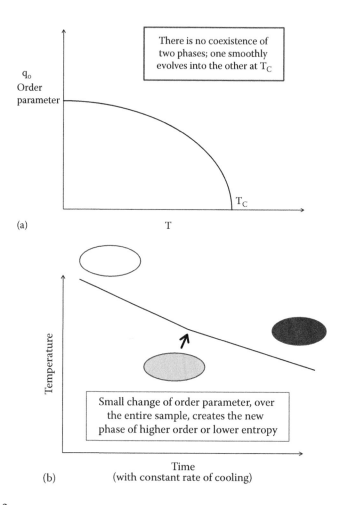

FIGURE 2.2
We show in schematic (a) the variation of order parameter q_o, as the control variable (temperature) is varied for a continuous phase transition. There is a smooth evolution and continuous
change in q_o at the transition temperature T_C. The two phases become indistinguishable at this
point, consistent with both having the same order parameter at T_C. As heat is removed, the temperature changes continuously, as is brought out in schematic (b) where we show the variation
of temperature under a constant cooling rate. The formation of a lower entropy (or higher order
parameter) phase on cooling occurs simultaneously throughout the sample as T_C is crossed, as
brought out by the schematic.

simultaneously over the entire sample volume, and the correlation length is infinite. The transition initiates with the onset of infinitesimal change in the order parameter but over large length scales.

As discussed in Chapter 1, the higher T phase always has larger entropy or disorder. Consequently, phase-2, which exists at $T > T_C$, has a lower order and is chosen to have an order parameter $q = 0$ while phase-1, which exists at $T < T_C$, is more ordered and has a finite order parameter, say $q_0(T)$. For a first order transition, $q(T_C)$ has the two allowed values: 0 and $q_0(T_C)$; for a continuous phase transition, it has only one allowed value: 0. The allowed state at each T must correspond to a minimum in free energy G. Instead of considering two distinct free energy values, G_1 and G_2, corresponding to the two phases, as in Chapter 1, and only checking which of the two is lower, we shall now consider $G(q, T)$ as a function of the order parameter q at each T and look for the minimum in G at each T. In Chapter 1, the lower of G_1 and G_2 corresponded to the equilibrium phase; the minimum in $G(q, T)$ is now occurring at the equilibrium value of $q(T)$. It follows that for a first order transition, $G(q, T_C)$ would have two minima, viz., at $q = 0$ and at $q = q_0(T_C)$ for two coexisting equilibrium phases. For a continuous phase transition, $G(q, T_C)$ would have only one minimum, viz., at $q = 0$. Now, are all the states with the continuously variable order parameter q physically realizable at each temperature, and do they have a physical significance? We note here that for a q to be physically realizable it must be a minimum in $G(q, T)$ at some T. The question of that q being physically realizable at some other T, where it does not correspond to such a minimum, shall become relevant when we discuss arrested states later in the book.

We have mentioned another difference between phase transitions involving and not involving latent heat. The former must take a finite time for removal of the finite latent heat, whereas temperature crosses T_C instantaneously in a continuous phase transition. The time for removing latent heat and crossing T_C in a first order transition is dictated by the time required for the order parameter to change, and the smallest volume over which the new phase nucleates has the length scale of the correlation length. This volume is finite, and the time is dictated by kinetics. This would depend on the temperature (kinetics is slower at lower temperatures) and pressure (kinetics is hindered at higher densities), and for a magnetic transition it could also depend on the magnetic field (magnetic moments would stay aligned at a higher magnitude of H). This time, over which nuclei of the new phase form, could sometimes even be much longer than the laboratory time scales on which measurements are made. We thus have another difference between first order and continuous phase transitions: A first order transition could be inhibited or arrested for some range of values of the control variable, whereas this is not possible for a continuous phase transition. The formation of a structural glass, resulting from the inhibition of the crystallization of a liquid, is an example of such an arrest of a first order transition. We shall discuss such arrested transitions in detail in Chapter 4.

We have enumerated two observational characteristics of first order transitions dictated by their having finite latent heat, namely, the two phases coexist at the transition point and the kinetics of the phase transition can be arrested. We assert that phase coexistence is essential in first order transitions but is absent in continuous phase transitions.

2.2 Metastable States Are Specific to First Order Transitions

In the Ehrenfest classification, there was no clear conclusion about the possibility of supercooling or superheating across second order or higher order transitions. Supercooling or superheating only required that the material should persist to follow the free-energy trajectory of the earlier phase, as indicated by the dashed lines in Figures 1.5 and 1.6. We shall understand in this section why such behavior is possible across first order phase transitions and why it is not possible to supercool or superheat across continuous phase transitions.

Let us consider a material undergoing a first order phase transition at T_C. The higher temperature/higher entropy phase-2 is disordered and, without loss of generality, has been assigned an order parameter value $q = 0$. Since this is the stable state at $T > T_C$, the free energy $G(q)$ must have a minimum at $q = 0$. All other values of q must correspond to states that have a higher value of G and are not stable. If a state with another value of q is metastable, then it would transform to the $q = 0$ state under a small disturbance but not an infinitesimal one. This would require $G(q)$ at that q to have a local minimum that would be shallow and separated from the absolute minimum by a barrier in $G(q)$. The height of the barrier would dictate the minimum disturbance required for the metastable state to transform into the stable state by going to the absolute minimum at $q = 0$. (The limitation of the Ehrenfest classification, stated as "Problem 3" in Section 1.6 can be addressed with this concept.) If there are no metastable states (say at T much higher than T_C) then $G(q)$ must have only one minimum at $q = 0$ and increase monotonically on either side. This is schematically depicted in Figure 2.3a.

At $T = T_C$, the two phases corresponding to $q = 0$ and $q = q_0(T_C)$ coexist as stable states. $G(q)$ must have two equal minima, since each of these two states is stable. These minima must be separated by a barrier that is essential to keep each phase from spontaneously converting to the other. This is depicted in Figure 2.3b. The low temperature phase is ordered and has a finite order parameter, which is q_0 at T_C, and the order parameter rises as T is lowered further. If all other values of q correspond to states that are not stable at $T \ll T_C$, then $G(q)$ must have a minimum at some finite q_0 and increase monotonically on either side. This is schematically depicted in Figure 2.3c.

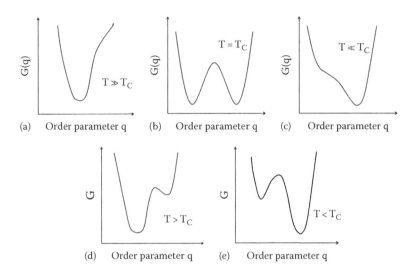

FIGURE 2.3
We plot the schematic G(q) as a function of q for a first order phase transition satisfying the criterion of coexisting phases at T_C. In (a), we show the schematic for $T \gg T_C$; in (b), we show the schematic for $T = T_C$; in (c), we show the schematic for $T \ll T_C$. The schematic in (d) considers a temperature intermediate between the cases (a) and (b), and the schematic in (e) considers a temperature intermediate between the cases (b) and (c). We note the existence of a metastable state in these cases.

We now discuss how G(q, T) evolves as T is lowered from the high value, corresponding to Figure 2.3a, to T_C and then to lower T. The change from 2.3a to 2.3b cannot be a sudden change at some temperature. This is also true of the change from 2.3b to 2.3c. At a temperature slightly above T_C, there must be a shallow minimum at finite q with a value of G slightly higher than $G(q = 0)$ to conform with q = 0 being the stable state. This is depicted in Figure 2.3d. We now have the possibility of two distinct phases existing at that T: one phase corresponding to the absolute minimum at q = 0 that is stable, and the other phase corresponding to the shallow local minimum at q close to $q_C(T_C)$ that is metastable. Similarly, at a temperature slightly below T_C, there must be a shallow minimum at q = 0 with a value of G slightly higher than $G(q = q_C(T))$. The absolute minimum is at a finite q, as depicted in Figure 2.3e. We, now, again have the possibility of two distinct phases existing at that T: one phase corresponding to the shallow local minimum at q = 0, which is metastable, and the other phase corresponding to the absolute minimum at q close to $q_C(T_C)$, which is stable. This existence of local minima corresponding to metastable states at T near T_C is a general result for all transitions where two phases coexist at T_C, i.e., for all first order transitions.

The scenario described by Figure 2.3a and c is true for all phase transitions, but the one described by Figure 2.3b, d, and e is true only for a first order phase transition at $T = T_C$ and corresponds to the schematic in Figure 2.1.

Thus, for the case of Figure 2.1, Figure 2.3e represents an essential inter-mediate state between 2.3b and c, and Figure 2.3d represents an essential intermediate state between 2.3a and c. *The existence of metastable states is, therefore, essential to first order phase transitions.* As a limitation of the Ehrenfest classification, we record that it did not conclude that metastable states are essential to first order transitions.

The schematic in Figure 2.2 does not envisage two phases coexisting at any temperature, and the scenario described by Figure 2.3b, d, and e will not be realized in such continuous transitions as an intermediate between Figure 2.3a and c.

In the case of a continuous phase transition, there is no latent heat and no coexistence of the two phases—both phases have the same order parameter ($q = 0$) at T_C. Phase-2 with $q = 0$ is the absolute minimum of $G(q)$ for all $T > T_C$, and this is also the order parameter at T_C where phase-2 and phase-1 are indistinguishable. However, phase-1 develops a nonzero order parameter when T is lowered even slightly below T_C. This is depicted in the schematic in Figure 2.4. This continuous increase in the order parameter $q_0(T)$ below T_C requires that there be no barrier at $q = 0$ and $G(q)$ should actually have an inflection point at $q = 0$ at T_C.

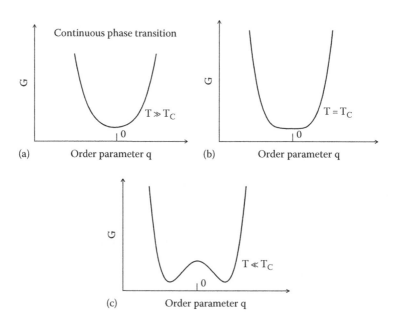

FIGURE 2.4
We plot the schematic $G(q)$ as a function of q for a continuous phase transition. In (a) we show the schematic for $T \gg T_C$, in (b) for $T = T_C$, and in (c) for $T \ll T_C$. The two phases become identical at T_C with an order parameter $q = 0$. The order parameter of the stable state, corresponding to the minimum in G, increases as T is lowered below T_C. There is one minimum for $T \ll T_C$; two minima are seen in (c) because there is an inversion symmetry between q and $-q$.

The evolution from a minimum to an inflection point with lowering T requires that d^2G/dq^2 at $q = 0$ should change continuously from being positive at $T > T_C$ to being zero and then being negative. This is depicted in Figure 2.4. There is no barrier in $G(q)$ at $q = 0$ as T drops below T_C, and $q = 0$ is not a local minimum for $T < T_C$. *A continuous phase transition does not permit the existence of a local minimum in G(q), and it does not permit the existence of metastable states.* We record another limitation of the Ehrenfest classification that it did not conclude that second (and higher) order phase transitions cannot have metastable states.

In the modern classification, phase coexistence at T_C and the existence of latent heat emerge as accepted characteristics distinguishing first order phase transitions from continuous phase transitions. We have argued earlier that metastable states can persist only across first order transitions, and that too only over a range of values of the control variable in the neighborhood of the phase transition. These general assertions on metastable states allow new signatures for identifying the type of phase transition. Supercooling and superheating are possible only across first order phase transitions, and hysteresis will be observed while increasing and decreasing temperature across T_C. The absence of metastable states in continuous phase transitions rules out the possibility of hysteresis. It follows that the transition under the increasing and decreasing values of a control parameter can show hysteresis, and hysteresis in the occurrence of the transition is possible only across first order phase transitions. *Hysteresis (or its absence) is thus a signature for identifying the type of phase transition.*

We caution that hysteresis can occur in a transition because of an experimental artifact such as thermal lag. Also, hysteresis can be seen without there being any underlying phase transition due to the slowing down of kinetics. Both these points become very important if we use hysteresis as a signature for a first order transition, and this will be discussed in Chapter 3.

We have developed the modern classification of phase transitions with an emphasis on metastable states. The definition of a first order transition in terms of latent heat is unchanged from the Ehrenfest classification, but the emphasis has been shifted and is now on phase coexistence at T_C and not on discontinuities in derivatives of free energy. As we proceed, we shall recognize a shift in emphasis away from latent heat and the Clausius–Clapeyron relation. This shift in emphasis is essential to the subsequent discussion, in Chapter 5, on the recent common observations in practically relevant magnetic materials. These materials are multicomponent compounds or have doping to enhance some of their physical properties; these provide quenched disorder that broadens the first order magnetic transitions. Discontinuities in the first derivatives of the free energy get broadened or smudged. Latent heat appears only as a peak in specific heat in broadened first order transitions and cannot be quantified to test the Clausius–Clapeyron relation. The shift in emphasis away from latent heat and from the Clausius–Clapeyron relation has become a necessity. *The emphasis will then be on the observation of hysteresis in the occurrence of the transition, subject to cross-checks essential for ruling out thermal lag and other artifacts.*

We now start an analysis of metastable states across first order transitions and will develop an understanding of how these metastablities will behave under different histories of control parameters.

2.3 Limits of Metastability

Figure 2.1 depicts two phases, with different values of the order parameter, coexisting as equilibrium states at T_C. As discussed earlier, the equilibrium state corresponds to the absolute or global minimum in G. Therefore, there must be two equal minima in G at T_C, with the minima occurring at $q = 0$ and at some $q = q_c$. This was depicted in Figure 2.3b. At very large T, the minimum will correspond to the phase with higher entropy, and, as depicted in Figure 2.3a, the order parameter $q = 0$ must then correspond to an absolute minimum of G(q). G(q) must always be bounded at large q, so G(q) must monotonically rise to infinity as q becomes large. At very low T ($\ll T^*$), the disordered higher entropy state with $q = 0$ must not be an allowed state, so $G(q = 0)$ cannot be a local minimum. We have depicted in Figure 2.3 G(q) as a function of temperature satisfying these generic rules. The simplest polynomial satisfying these rules is given by

$$G(q) = a(T - T^*)q^2 - bq^3 + cq^4 \tag{2.1}$$

where each of a, b, and c are positive [1,2]. The condition of two equal minima (at $q = 0$ and at $q = q_c$) at $T = T_C$ gives the following two equations:

$$G(q) = 0 \tag{2.2}$$

$$dG/dq = 2a(T_C - T^*)q - 3bq^2 + 4cq^3 = 0 \tag{2.3}$$

These must be satisfied at both $q = 0$ and at $q = q_c$. Since these values must correspond to minima in G(q), we shall further require that at both these values of q we must have

$$d^2G/dq^2 = 2a(T - T^*) - 6bq + 12cq^2 > 0 \tag{2.4}$$

Both Equations 2.2 and 2.3 are obviously satisfied at $q = 0$. Further for $T > T^*$, $d^2G/dq^2 = a(T - T^*)$ at $q = 0$ is also positive, ensuring that $q = 0$ corresponds to a minimum in G(q) for all T > T*. We note that q = 0 is an inflection point at T = T* and a local maximum at T < T*. T* gives the limiting temperature below which the q = 0 state becomes unstable. We now check for the second minimum, at $q = q_C(T)$.

Substituting from (2.1) in Equation 2.2 at $q = q_C$, we get that

$$a(T_C - T^*) - bq_C + cq_C^2 = 0 \qquad (2.5)$$

From Equation 2.3 at $q = q_C$, we get

$$2a(T_C - T^*) - 3bq_C + 4cq_C^2 = 0 \qquad (2.6)$$

From Equations 2.5 and 2.6, we get

$$-bq_C + 2cq_C^2 = 0, \quad \text{or}$$

$$q_C = b/2c \qquad (2.7)$$

Substituting back in Equation 2.5, we get

$$a(T_C - T^*) - b^2/2c + cb^2/4c^2 = 0, \quad \text{or}$$

$$T_C = T^* + b^2/(4ac) \qquad (2.8)$$

Substituting $q = q_C = b/2c$ in (2.4), we get d^2G/dq^2 $(q = q_C, T = T_C) = 2a(T_C - T^*)$, and from (2.8), we get d^2G/dq^2 $(q = q_C, T = T_C) = b^2/2c$ and that it is positive, confirming that $q = q_C = b/2c$ is a minimum in G(q). We thus have two equal minima at T = T_C, at q = 0 and at $q_C = b/2c$. To find a maximum separating these two minima at T_C, we substitute (2.8) in Equation 2.3 and find a third solution for dG/dq = 0, viz., at $q_B = b/4c$. This does correspond to a maximum in G(q) because, from Equation 2.4, we get at this $q_B = b/4c$ that d^2G/dq^2 $(q = q_C, T = T_C) = -b^2/4c < 0$. We thus have two equal minima at T = T_C that are separated by a barrier. The height of the barrier is obtained using Equation 2.1 as follows:

$$G(q = q_B, T = T_C) = b^4/256c^3$$

This is shown in the schematic in Figure 2.3b.

Further, we note that for T < T_C the state at $q = q_C$ must correspond to the absolute minimum, while for T > T_C the state at q = 0 must also correspond to the absolute minimum. Detailed plots of G(q, T) given by Equation 2.1 can be obtained for various T, as are depicted in the schematic Figure 2.3.

We make some generic observations at this stage, which are valid for other functional forms of $G(q)$ that would show a first order phase transition [1].

For $T < T_C$, the minimum at $q = 0$ is a local minimum, corresponding to a supercooled higher entropy phase. The barrier in $G(q)$ separating it from the absolute minimum at finite q_0 decreases continuously as T falls and moves to smaller values of q, making the minimum corresponding to the supercooled state shallower and less stable. As the barrier vanishes at $T = T^*$, the local minimum at $q = 0$ becomes an inflection point in $G(q)$ and the absolute minimum corresponding to the lower entropy state is at $q_C = 3b/(4c)$. Similarly, for $T > T_C$, the minimum at finite q_C is a local minimum, corresponding to a superheated lower entropy phase. The barrier in $G(q)$ separating it from the absolute minimum at $q = 0$ decreases as T rises above T_C and moves to larger values of q, making the superheated state less stable. This barrier vanishes at T^{**} when the superheated becomes unstable. The barrier separating the metastable and the stable states is largest close to T_C and reduces continuously as T moves toward either limit of metastability, and the conversion from the metastable to stable state requires reducing energy. These generic statements are valid for all first order transitions.

As the barrier around the local minimum in $G(q)$ vanishes at T^* or at T^{**}, this local minimum becomes an inflection point in $G(q)$. At these limiting temperatures, the metastable state becomes unstable under infinitesimal fluctuations, and we have a prescription for estimating these limiting temperatures beyond which the metastable states cannot persist. Such limits were obtained for the specific case of the van der Waals model, but there was no general prescription in the Ehrenfest classification.

The parametric temperature T^* corresponds to $d^2G/dq^2 = 2a(T - T^*) - 6bq + 12cq^2 = 0$ at $q = 0$. Thus, $G(q)$ develops a point of inflection at $q = 0$ at $T = T^*$. $G(q) = 0$ and $dG/dq = 2a(T - T^*)q - 3bq^2 + 4cq^3 = 0$ at $q = 0$ at all temperatures and so also at $T = T^*$. The point $q = 0$, thus, evolves from a local minimum into an inflection point, completing the transformation of an equilibrium state above T_C to a metastable supercooled state below T_C and, as T is lowered further, to an unstable state at T^*. We have, thus, found a limit to supercooling at T^*. For $T < T^*$, $d^2G/dq^2 < 0$ at $q = 0$, and $q = 0$ is now a local maximum and the supercooled state has become unstable. The limit of metastability is generic to all first order transitions, and this has followed seamlessly from the requirement that the two phases, with different order parameters, must coexist at T_C.

As T is raised above T_C, the global minimum at q_C evolves into a local minimum with $q = 0$ corresponding to the global minimum, and its position moves to lower q. As T is raised further, the barrier surrounding the local minimum reduces in height and the local minimum becomes shallower. Finally, at some $T = T^{**}$, the barrier becomes 0, the local minimum becomes an inflection point, and the superheated lower entropy phase becomes unstable. This is obtained again as a solution to the two equations $d^2G/dq^2 = 2a(T^{**} - T^*) - 6bq + 12cq^2 = 0$ and $dG/dq = 2a(T^{**} - T^*)q - 3bq^2 + 4cq^3 = 0$, which are solved for T^{**} and $q = q_C$ but at finite q. We eliminate T^{**} by simply subtracting the two equations and get $q_C = 3b/8c$. Substituting this back, we get

$$2a(T^{**} - T^*) - 3b(3b/8c) + 4c(3b/8c)^2 = 0, \quad \text{or}$$

$$T^{**} - T^* = 9b^2/32ac$$

which gives us the following:

$$T^{**} = T_C + b^2/32ac \qquad (2.9)$$

We have now found a limit to superheating, along with the order parameter ($q_C = 3b/8c$ at T^{**}) of the superheated state at this limit.

Finally, the entropy jump at the equilibrium transition at T_C is also obtained by taking the difference of $(-\partial G/\partial T)$ at $q = 0$ and at $q = q_C$ ($=b/2c$ at $T = T_C$), and we get

$$\Delta S = \left(-\partial G/\partial T\right)_{(q=0)} - \left(-\partial G/\partial T\right)_{(q=qc)} = 0 - aq_C^2 = ab^2/4c^2 \qquad (2.10)$$

Before concluding this discussion, we shall consider some free energy expansions for continuous phase transitions. Equation 2.1 represents a continuous phase transition under the condition $b = 0$ since ΔS vanishes, and we also get $q_C = 0$ at $T = T_C$. We then get

$$G(q) = a(T - T^*)q^2 + cq^4 \quad \text{and}$$

$$G(q) = 0 \quad \text{at } q = 0 \text{ for } T \geq T^*$$

For $T \leq T^*$, we have $G(q_C) = 0$ at $q_C = [(a/2c)(T^* - T)]^{1/2}$.

It follows that $T_C = T^*$ and that the order parameter corresponding to the minimum in G(q) rises smoothly as T drops below T_C. There is always only one minimum in G(q) and no local minimum, so there are no metastable states and there is no possibility of supercooling or superheating. Thus, there cannot be any hysteresis across a continuous phase transition. We shall not consider these continuous phase transitions further.

2.4 Hysteresis across First Order Phase Transitions

In the van der Waals model, we found limits of metastability as the points till where we had multiple real solutions satisfying the physical constraint that compressibility was not negative. That was for a specific model. Here, we have shown limits determined by the point where the local minimum in free energy becomes an inflection point. The treatment here is generic to all first order transitions and has a general applicability. More importantly, the van der Waals model did not provide any estimate of the energy (or fluctuations) required to convert a metastable state to the stable state; however, here, we can estimate the free energy barrier $G(q_B)$ that needs to be crossed to convert the metastable state to the stable state. In the van der Waals model, we showed that the limits to metastability exist for temperature decreasing and increasing, as well as for pressure increasing and decreasing, with the limits being the same, irrespective of the control variable used. On the other hand, we have so far shown this metastability for the temperature increasing and decreasing case only; we shall extend this to other control variables later in this chapter.

At this stage, we wish to concentrate on the significance of Equations 2.8 and 2.9, but the general applicability of supercooling and superheating limits ensures that corresponding results can be found for any functional form of $G(q)$ that has two equal minima at $T = T_C$. As an exercise, we choose

$$G(q) = a(T - T^*)q^2 - bq^4 + cq^6 \qquad (2.11)$$

Following steps similar to those following Equation 2.1, we get

$$T_C = T^* + b^2/(4ac) \qquad (2.12)$$

The two equal minima at $T = T_C$ are obtained at $q = 0$ and at $q_C = [b/(2c)]^{1/2}$, with a barrier at $q_B = [b/(6c)]^{1/2}$ of barrier height $G(q_B) = b^3/54c^2$. The latent heat at T_C is $L = abT_C/(2c)$. As earlier, T^* is the limiting temperature where $q = 0$ changes from being a minimum in $G(q)$ to being an inflection point and is the limiting temperature to which the higher entropy phase with $q = 0$ can be supercooled. The limiting temperature to which the lower entropy phase can be superheated is similarly obtained for this case as follows:

$$T^{**} = T^* + b^2/(3ac) = T_C + b^2/(12ac) \qquad (2.13)$$

We, thus, again find that when two phases coexist at the transition temperature, then supercooling and superheating are possible across the transition temperature as the global minimum becomes a local minimum in free energy.

The metastable phase can persist as the local minimum becomes shallower, up to limiting temperatures where the local minimum finally evolves into an inflection point. This follows intuitively from the schematics in Figure 2.3 and is a universal result for all first order transitions.

The discussion on the van der Waals model concluded that supercooling and superheating are possible but did not lead to any conclusion on the conditions under which this will or will not be observed. Here, we have a visualization of a free energy barrier $G(q_B)$, which is a function of T and decreases monotonically as T moves away from T_C. This barrier separates the absolute minimum from the local minimum, and the metastable state in the local minimum can transform to the stable state if it can cross the barrier, i.e., if it possesses fluctuations of energy E_f large enough to satisfy

$$E_f + kT \sim G(q_B) \tag{2.14}$$

We assume here that the system is able to explore the phase space corresponding to different values of q on the experimental time scale and, in that sense, the local minima in $G(q)$ at various T are also thermodynamic states even though they are not equilibrium states. When we cannot explore this phase space, then the lack of dynamics will preclude thermodynamics, and the values of q that do not correspond to even the local minima at that T become relevant to experimental observations. We shall come back to this in Chapter 4.

Experimental studies have driven new concepts, and studies on supercooling of liquids below the freezing point have been abundant. We list in Table 2.1 the melting points and the limits to which supercooling has been observed for some metallic elements and for the most widely studied and common compound—water. The values of T* vary across experiments, mainly because the limit to which supercooling can be observed is dictated by parameters such as purity of the liquid and cleanliness of the container surface, which can contribute to E_f in Equation 2.14. We note that most liquids can be supercooled to around 0.8–$0.85T_m$, but data on superheating of solids is scarce. Some understanding of this follows from the experimental technique of Itami et al. [3], whose results have been used to generate much of the data in Table 2.1. They used a containerless technique to avoid nucleation at inhomogeneities on container surfaces. They also divided the liquid metal into small droplets so that impurities were restricted to only some droplets while others were pure and could supercool. By similar arguments, it is difficult to superheat solids because the surface provides sites that favor liquid nucleation, or melting. Surfaces add to E_f and preclude superheating of solids.

Our treatment of the modern classification of phase transitions emphasizes some results of the van der Waals model as general to all first order phase transitions. Supercooling and superheating are possible, and they will occur if E_f is small such that the relation (2.14) is not satisfied.

TABLE 2.1

Melting Points T_m, and the Supercooling Limits T^*

	T_m (K)	T^* (K)
Ruthenium	2607	2180
Rhodium	2237	1810
Hafnium	2506	2175
Tantalum	3290	2590
Tungsten	3695	3100
Rhenium	3459	2660
Iridium	2719	2280
Gallium	303	245
Water	273	233
Aluminum	932	772
Cadmium	594	484
Copper	1356	1120
Iron	1803	1508
Gold	1336	1106
Platinum	2043	1673
Silver	1233	1007
Tin	506	319

Source: Using data from Itami, T. et al., *Mater. Trans.*, 51, 1510, 2010.

The transition on cooling occurs at $T \le T_C$, and the transition on warming occurs at $T \ge T_C$. If either supercooling or superheating occurs, then the transition on cooling will be at a temperature lower than that observed on heating and we shall observe hysteresis as temperature is lowered and raised. We have also understood that hysteresis cannot occur across continuous phase transitions. Hysteresis in the transition temperature is, accordingly, a sufficient condition to identify a phase transition as first order. The van der Waals model, in addition, allowed thermal hysteresis with pressure as the control variable. We shall discuss hysteresis with control variables other than temperature, in this understanding of first order transitions, in the following sections.

2.5 Metastable to Stable Transformations

We have understood that between T^* and T^{**} the free energy $G(q)$ has two minima. One is a local minimum and corresponds to a metastable state. The other is a global or absolute minimum and corresponds to the stable state. Further, there is a history dependence that dictates whether the system is in the metastable or stable state, as explained in the schematic in Figure 2.5.

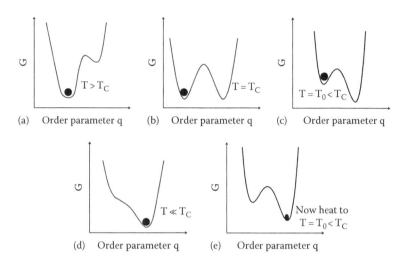

FIGURE 2.5

We show a schematic of the state at a temperature T_0 lying between the supercooling and superheating limits depending on the thermal history. The occupied state is depicted by the filled circle. We cool from $T \gg T_C$ in (a), then to T in (b), and then to below T_C in (c) but remain above T^*, and we are in the supercooled metastable state at T_0. We now cool to below T^* as depicted in (d) and then warm to the same T_0 as depicted in (e). On following this path, we are in the equilibrium stable state at T_0. Such a path-dependence of the state is possible for T_0 in the temperature window between T^* and T^{**}.

The material can be in a metastable supercooled state below T_C if it is cooled from above T^{**} but has not been cooled to a temperature below T^*, and it can be in a metastable superheated state if it has been warmed from below T^* but has not been heated to a temperature above T^{**}. If it has been cooled below T^*, then it must have made a transition to the lower entropy phase-1, and, similarly, if it has been heated above T^{**}, then it must be in the higher entropy phase-2. We shall see that variations in the second control variable make the history dependence more complicated.

The earlier discussion has also brought out that the barrier $G(q_B)$, separating the absolute minimum from the local minimum, decreases monotonically as T moves away from T_C. Assuming that the system is able to explore the phase space corresponding to different values of q on the experimental time scale, the metastable to stable transformation will occur when the barrier decreases such that the relation (2.14) is satisfied. This can happen by either $G(q_B)$ being reduced as T approaches T^* (or T^{**}) while decreasing (or increasing) temperature—as discussed in the preceding paragraph—or E_f being raised. The former can be checked by studying the relaxation rates at different temperatures, while the second option requires appropriate experimental protocols. We first consider the methods of increasing E_f.

Users of microwave ovens are routinely warned about burns or scalding injuries to hands and face if water is heated in a clean cup. This is because it can result in superheated water if the cup is clean and its surface is smooth. Superheating is prevented if foreign materials such as instant coffee or sugar or salt are added before heating. These impurities would act as centers for nucleating the gas phase, as mentioned in the preceding section. However, once superheated water has formed, a slight disturbance or movement such as picking up the cup would result in the metastable superheated water converting to the equilibrium stable state of steam. This sudden conversion of water to steam causes a violent eruption, with the boiling water exploding out of the cup and causing burns or scalding injuries. We, thus, see that slight movements to the cup add to E_f, and if the superheated water is far from T_C (which is the boiling point in this case) $G(q_B)$ would be small and this slight addition to E_f is enough to cross the barrier. It follows that constant stirring would provide a large E_f and is an experimental addition to ensure that the boiling of water is observed at the standard boiling point. We find that a superheated metastable liquid can be converted to the stable gas phase by a physical disturbance that adds to E_f.

Let us now consider a supercooled liquid. Again, inhomogeneities on the container surface and impurities in the liquid provide nucleation sites for freezing the solid. Water can be supercooled relatively easily to temperatures slightly below its melting point, when nucleation sites will initiate freezing. Careful experiments allow supercooling to a limiting temperature of 233 K before freezing starts. There are movie clips available on the Internet showing that if one takes a bottle of water that has been supercooled and taps it at the bottom, then some of the water freezes instantaneously. An old folklore was that a pinprick will freeze supercooled water. Hunt and Jackson [4] reported, in 1966, rigorous experiments to establish that "agitation induces crystallization." They reported that stirred water allows only about 0.5°C supercooling below the equilibrium freezing point before spontaneous nucleation and freezing, whereas undisturbed water allows over 5°C supercooling under similar conditions. Agitation, or stirring, increases E_f. The barrier $G(q_B)$ is higher when T is closer to T_C, and stirring or agitation helps the metastable supercooled liquid to transform to the stable solid when T is closer to T_C.

We now discuss experiments following studies on the vortex liquid to vortex solid transition discussed in Section 1.2. This transition was first theoretically conceptualized in high-T_C superconductors, because of the high temperature at which superconductivity persists and the small coherence length. The latter restricts the number of Cooper pairs in a coherence volume, while the former allows fluctuations that would be more prominent in this small coherence volume. As discussed in Chapter 1, this melting transition was firmly and rigorously established in both $Bi_2Sr_2CaCu_2O_8$ (BSCCO) and $YBa_2Cu_3O_7$ (YBCO), with the observation of latent heat and confirming compliance with the Clausius–Clapeyron relation. The concept of vortex-lattice melting would hold even in conventional superconductors like niobium, though the signature would be less prominent because of its lower T_C and larger correlation length and would require more sensitive experiments.

Measurements of latent heat for vortex-lattice melting have not been reported in conventional superconductors, and no tests for Clausius–Clapeyron relation are known, though, as we shall discuss in Chapter 3, the melting of the vortex lattice in niobium has been established as a first order transition [5]. In this case, as in the case of charge-order melting in half-doped manganites, the existence of a first order transition could be argued through the observation of hysteresis without the rigor of establishing latent heat and without checking for compliance with the Clausius–Clapeyron relation.

Chaddah and Roy [1] discussed the behavior of metastable states and hysteresis under the variation of two control variables. They discussed how E_f at a specific (P, T) could depend on the path followed in the space of two control variables and made specific predictions in the context of vortex matter. They argued that change in vortex density causes increase in E_f because the viscous motion of vortices results in dissipation of energy. They argued further that an oscillating magnetic field caused oscillations in vortex density and the dissipation this produces results in a higher E_f, causing the metastable state to transform to the stable state. As a corollary, the same oscillating magnetic field will not cause any transformation in the stable state since this already corresponds to an absolute minimum in G(q). The existence of a first order transition could be argued through the observation of hysteresis and its path-dependence in the space of two control variables, without the rigor of establishing latent heat and without checking for compliance with the Clausius–Clapeyron relation. Chaddah and Roy [1] argued that there was an inequality in the path-dependence of hysteresis associated with a first order transition: a path that involved variations in pressure or magnetic field would show less hysteresis than a path that only involved variations in temperature. They thus claimed a cross-check for hysteresis due to supercooling and superheating. In all these studies, where a first order transition has been established without a direct measurement of latent heat, the use of two control variables to traverse the phase transition is an underlying theme. We shall now address this.

2.6 First Order Phase Transitions with Two Control Variables

We noted in Chapter 1 that the Ehrenfest classification treats the two control variables—temperature and pressure—at par. The same was shown in Chapter 1 to be true for the van der Waals model, but the discussion in this chapter has followed common practice on the modern classification that treats only temperature as the control variable [2]. Let us now include a second control variable.

The second control variable is usually taken as pressure since all structural transitions are affected by pressure. Recently, electric-field-driven and magnetic-field-driven transitions have gained importance because of the potential applications of dielectric and magnetic materials in various devices. As a second control variable in conjunction with temperature, both magnetic and electric fields provide experimental ease when compared to pressure. This is because pressure, like temperature, is controlled through material contact, and the two control variables thus interfere with each other. This difficulty in controlling T while varying P is exemplified by the fact that measurements under isothermal variation of pressure, at temperatures well below ambient, are scarce. In contrast, magnetic field is transmitted and controlled through vacuum. Its variation does not interfere with the control of temperature, and isothermal variation of magnetic field at temperatures well below ambient is particularly commonplace. In fact, any protocol of alternate temperature and magnetic field variations that is conceived at temperatures of ambient and below can be experimentally executed within the range of magnetic fields available with superconducting magnets. All gedanken experiments involving variation of H and T are practically possible.

Magnetic first order transitions have application possibilities through the phenomena of magnetoresistance, magnetocaloric effect, magnetic shape-memory effect, exchange bias, and magneto-dielectric effect. Mainly because of such possibilities and the consequent growth of corresponding experiments, magnetic field as a control variable has now become common to commercial instruments operating at low temperatures. High magnetic fields are obtained using superconducting magnets, which require a cryogenic liquid helium environment for their operation. Thus, low temperature and high magnetic fields (LTHM) require a cryogenic environment around the sample. First order magnetic transitions at low temperatures have now been studied extensively for about three decades and have resulted in many new experimental results. In Chapter 1, we have noted K. Huang's statement, "Thermodynamics is a phenomenological theory of matter. As such, it draws its concepts directly from experiments" [6]. Many of the new concepts that form the *raison d'être* of this book have come from LTHM experiments. These results shall be brought out in subsequent chapters. Here, we shall restrict ourselves to extending the use of G(q) with a second control parameter other than T. We shall use the notation p for the second control variable, with the understanding that p can represent pressure P, magnetic field H, electric field E, or any other control variable whose variation changes T_C.

Since variation of the second control variable denoted by p changes T_C, it should also change T* in Equation 2.1, which we now rewrite as follows:

$$G(q) = a\left[T - T^*(p)\right]q^2 - bq^3 + cq^4 \qquad (2.15)$$

Following the arguments of Section 2.3, we get the limits of supercooling and superheating as

$$T^*(p) = T_C(p) - b^2/4ac \quad \text{and}$$

$$T^{**}(p) = T_C(p) + b^2/32ac$$

The maximum thermal hysteresis possible is

$$\Delta T = T^{**}(p) - T^*(p) = 9b^2/32ac \tag{2.16}$$

While each of T_C, T^*, and T^{**} depends on the value of the second control variable p, the extent of possible hysteresis given by Equation 2.16 now becomes independent of p. This appears unphysical. It was, accordingly, proposed by Chaddah and Roy [1] that the coefficient of q^2 should incorporate a further p-dependence in the parameter a. We, thus, rewrite Equation 2.15 as follows:

$$G(q, p, T) = a(p)\left[T - T^*(p)\right]q^2 - bq^3 + cq^4 \tag{2.17}$$

We now have $T^*(p) = T_C(p) - b^2/[4a(p)c]$ and $T^{**}(p) = T_C(p) + b^2/[32a(p)c]$. This introduces p-dependence in the extent of possible supercooling, the extent of possible superheating, and the possible hysteresis. In Figure 2.6, we show schematically variations of T^*, T_C, and T^{**} with the second variable p. In Figure 2.6a, we assume that T_C decreases when the second control variable rises. With pressure being the second control variable, this happens if the low temperature/lower entropy phase (phase-1) has higher volume than the high temperature/higher entropy phase (phase-2). This situation does not occur for the liquid-to-gas first order transition, but it does occur for many liquid-to-solid first order transitions, the most common being the water–ice transition. This is also seen, among others, in the case of bismuth, gallium, cerium, silicon, and germanium. When the second control variable is magnetic field, then this behavior is seen when the lower entropy phase is antiferromagnetic and the higher entropy phase is ferromagnetic. There are many such examples, and some will be discussed in later chapters. This is also seen in the normal-superconducting transition, where the lower temperature superconducting phase has lower magnetization. In Figure 2.6b, we assume that T_C increases when the second control variable rises. With pressure being the second control variable, this happens whenever the lower entropy phase has higher density than that of the higher entropy phase. This situation occurs for all physical liquid-to-gas and for most liquid-to-solid first order transitions. When the second control variable is magnetic field, then this behavior is seen when the lower entropy phase is a ferromagnet and the higher entropy phase is a paramagnet or an antiferromagnet. There are many such examples, as will be discussed in later chapters.

In each case, the $T^{**}(p)$ line derives from a free energy vs order parameter curve G(q, p), with a minimum at q = 0 and an inflection point at higher q corresponding to the superheated ordered state. The $T_C(p)$ line derives from a G(q, p) with two equal minima, at q = 0 and at a higher value q_0.

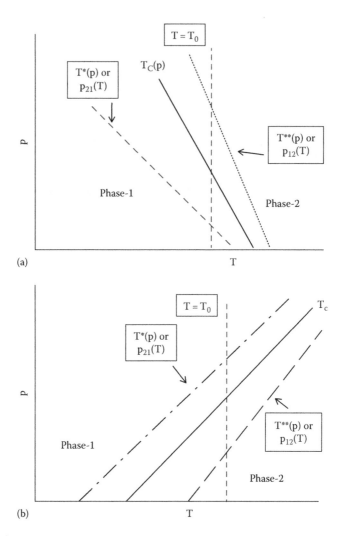

FIGURE 2.6
We show the schematic variation of T*, T_C, and T** with the second control variable. In (a) we assume that T_C decreases when the second control variable rises, and in (b) we assume that T_C increases when the second control variable rises. The schematic phase diagrams continue to make the simplistic assumption that these (T, p) lines are straight lines, though they are generally curved with the bending toward the temperature axis. The vertical dashed line in both figures corresponds to the isothermal variation of the second control variable crossing the limits of metastability at $p_{12}(T_0)$ and at $p_{21}(T_0)$.

The T*(p) line derives from a G(q, p) with a minimum at a finite q_0 and an inflection point at q = 0. This result follows from Equation 2.17 at various constant values of p and is true for all values of p. It follows that if p is varied with fixed T, then we will successively cross the lines T**(p), T_C(p), and T*(p). This result must also follow by solving G(q, p, T) with T fixed and p varied. While in the case of Figure 2.6a the limit of metastability with reducing p coincides with T*, in the case of Figure 2.6b it coincides with T**. This raises an issue of nomenclature, which we shall address shortly.

Both Equations 2.1 and 2.15 have been written by fixing the otherwise arbitrary origin of G(q) to satisfy G(0) = 0. In all standard discussions, this is assumed for all T and corresponds to the entropy for the disordered phase being chosen as 0 so that the free energy is independent of temperature at a fixed pressure. This is not obvious for Equation 2.17 when the parameter that is being varied is pressure and not temperature. As we increase the uniform pressure, free energy must increase, and this must be incorporated in the general case by adding a term $E_0(p)$ [1].

We now incorporate such a term and examine the behavior of the following:

$$G(q,p,T) = E_0(p) + a(p)\left[T - T^*(p)\right]q^2 - bq^3 + cq^4 \qquad (2.18)$$

For a fixed p, this reduces to Equation 2.1 and all that has been discussed earlier stands. For fixed p, we can assume that G(q = 0) = 0 at various T, but for fixed T, we cannot assume G(q = 0) = 0 at various p. G(q = 0, p), unlike G(q = 0, T), is not pegged at 0 but varies with p. Let us compare G(q, p, T) for various p but fixed T. We assume that T_C falls as p rises, and so do T* and T**. We choose a value of $T = T_0$ such that $T_0 < T^*$ at the lowest $p = p_0$. There is only one minimum in G(q) and there is no local minimum, and the higher entropy phase cannot exist as a metastable state. As we raise p, we have $T_0 = T^*(p^*)$, and this is the limiting pressure at which a local minimum exists for the higher entropy phase; q = 0 is an inflection point as depicted in Figure 2.7a, and this is the limiting pressure to which the supercooled state can exist as a metastable state. On further raising p, we encounter $T_0 = T_C(p_c)$, which is the pressure at which the two phases have the same free energy and where the phase transition would take place in the presence of large fluctuation energy E_f. At $p = p_c$, G(q) has two equal minima, at q = 0 and at $q = q_0$, as depicted in Figure 2.7b. If we raise p further, we get $T_0 = T^{**}(p^{**})$, which is the pressure at which the local minimum corresponding to the lower entropy phase becomes an inflection point, as in Figure 2.7c, and is the limit of metastability of the superheated state. At higher pressures, G(q) has only one minimum corresponding to the higher entropy phase, as in Figure 2.7d. At the highest pressure, there is only one minimum in G(q) and the material is in the disordered phase-2. The schematics in Figure 2.7e through 2.7h show that as p is lowered, we cross p_C and the material can remain in phase-2 as a metastable supercooled state. It becomes unstable when the limit of supercooling is reached at p* (now denoted by p_{21}). The difference between Figures 2.3 and 2.7 is that G(q = 0) is fixed at 0 in the former but is rising as p is rising in the latter.

In the schematic Figure 2.8a, we now consider the case where T_C rises as p rises. This is the case for all liquid–gas transitions, with the variable p being pressure, as also for a majority of solid–liquid transitions. This scenario also

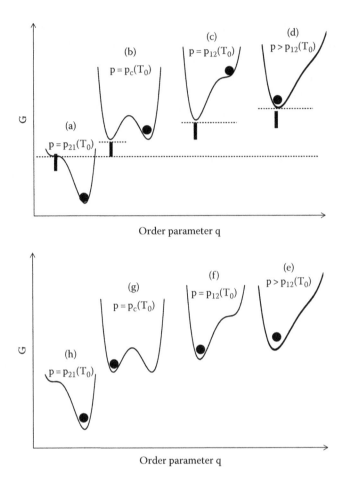

FIGURE 2.7
We plot the schematic G(q) as a function of q for various values of the second variable, for the case corresponding to Figure 2.6a. The temperature is held fixed at T_0. The plots correspond to different values of p and are staggered on the order parameter q-axis, but there is a broad vertical marker indicating the q = 0 disordered state for each plot. The vertical position for G(q = 0) would be 0 when the plots are for fixed p but varying T; here, the schematics show that it rises with increasing values of p for fixed T. Plots (a–d) are for increasing p with the solid circle depicting the state, while plots (e–h) are for decreasing p. The occupied state is depicted by the filled circle. There is a hysteresis with the control variable p, similar to that with the control variable T depicted in Figure 2.5, for p between $p_{21}(T_0)$ and $p_{12}(T_0)$.

corresponds, with the variable p being magnetic field, to various magnetic materials that undergo an antiferromagnetic to ferromagnetic transition with reducing T at zero field. Here we choose $T_1 > T^{**}$ at the lowest p, and we raise p till $T_1 < T^*$. As we raise p, we encounter $T_1 = T^{**}(p_{12})$. This is the limiting pressure at which the minimum corresponding to the superheated state becomes an inflection point and the lower entropy phase reaches its limit of

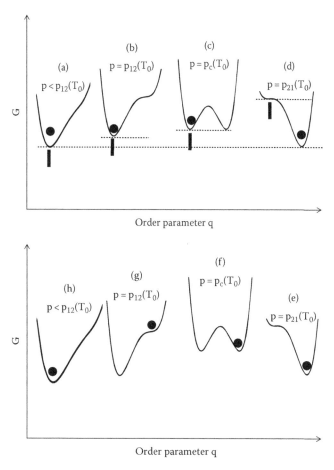

FIGURE 2.8
We plot the schematic $G(q)$ as a function of q for various values of the second variable, for the case corresponding to Figure 2.6b. The temperature is held fixed at T_0. The plots correspond to different values of p and are staggered on the order parameter q-axis, but there is a broad vertical marker indicating the q = 0 disordered state for each plot. The vertical position for $G(q = 0)$ would again be 0 when the plots are for fixed p but varying T; here, the schematics show that it rises with increasing values of p for fixed T. Plots (a–d) are for increasing p with the solid circle depicting the state, while plots (e–h) are for decreasing p. The occupied state is depicted by the filled circle. We again note a hysteresis for p between $p_{12}(T_0)$ and $p_{21}(T_0)$.

metastability on the reducing-p cycle. On further raising p, we encounter $T_1 = T_C(p_C)$, which is the pressure at which the two phases have the same free energy and the transition to phase-1 would take place in the presence of large fluctuation energy E_f. If we raise p even further, we encounter $T_1 = T^*(p_{21})$, which is the pressure at which the minimum corresponding to the supercooled state becomes an inflection point and the higher entropy phase reaches its limit of metastability on the increasing-p cycle. The schematics in (e)–(h) show the evolution of the state as p is lowered.

We have made our notation more generic and denote by p_{21}, p_{12} the pressures at which phase-2 and phase-1, respectively, become unstable. These represent, respectively, the limits to which the higher entropy phase-2 and the lower entropy phase-1 can remain metastable. Similar limits for the other control variable would be H_{21} and H_{12}.

If, however, T_0 is such that $T_0 > T^*$ till $p = 0$ for Figure 2.6a, then the metastable to stable transformation may not take place. Isothermal variations in p can result in irreversibilities in the remnant state. We show this in the schematics in Figure 2.9a, where we consider isothermal variations of p, from 0 to a large value well above p_{12} (earlier denoted by p^{**}) and back to 0, at various T_1. If $T_1 < T^*(p = 0)$, then both the starting and the final states are phase-1. If $T_1 > T^{**}(p = 0)$, then both the states are phase-2. If T_1 is above T_C but below T^{**}, then the starting state can be phase-2 if T was not lowered below T_C and the superheated phase-1 if T was lowered below $T^*(p = 0)$, while the final state will be phase-2. The initial and final states are different if T_1 is reached from a low temperature. If T_1 is below T_C but above T^*, then the starting state can be the supercooled phase-2 if T was not lowered below T^* and phase-1 otherwise, while the final state is supercooled phase-2. Again, the initial and final states are different if T_1 is reached from a low temperature. To the experimental observer, the significant conclusion from all these cases is that the

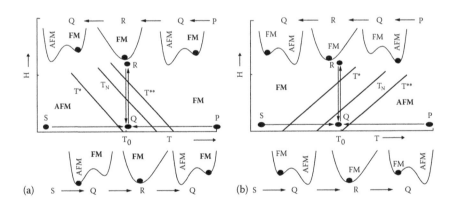

FIGURE 2.9

We show the hysteretic behavior on varying the second control variable with $T = T_0$ in the window between T^* and T^{**} where the initial state, as depicted in Figure 2.5, depends on whether T_0 is reached by cooling (from above T^{**}) or by heating (from below T^*). Such variations of the second control variable at arbitrary values of T are experimentally feasible when the second control variable is magnetic field (and difficult when it is pressure). While the phenomenology holds for *any* second control variable, such observations were performed for first order magnetic transitions between ferromagnetic (FM) and anti-ferromagnetic (AFM) states. This conceptual schematic, with T_N denoting the transition temperature. The occupied state is depicted by the filled circle. It shows that the hysteresis on varying H in this range of T_0 results in an open hysteresis loop, but for only one path of reaching T_0. In case (a) this happens when we heat from below T^*, and in case (b) this happens when we cool from above T^{**}. (Taken from Kushwaha, P. et al., *J. Phys. Cond. Matter.*, 20, 022204, 2008.)

initial and final states are different only for T_1 lying between $T^*(p = 0)$ and $T^{**}(p = 0)$, and then also only when T_1 is reached by warming from a low T and not when it is reached by cooling from a high T. Experimental verification requires isothermal variations over a large range of p at various values of T. As mentioned earlier, this poses difficulties when p is pressure, and no experimental verification has been reported. Nevertheless, this is a common experimental protocol with the generic control variable p being magnetic field, and experimental verifications have been reported in recent years.

We now consider the case depicted in Figure 2.9b, and the argument follows that for Figure 2.9a. It is left as an exercise. In this case the significant experimental conclusion from all cases is that the initial and final states are different only for T_1 lying between $T^*(p = 0)$ and $T^{**}(p = 0)$, and then also only when T_1 is reached by cooling from a high T and not when it is reached by warming from a low T. This is depicted in Figure 2.9b.

One example of a measurement depicting this behavior is shown in Figure 2.10. Kushwaha et al. [7] studied a magnetic first order transition (the generic second control variable p is now magnetic field) where the lower entropy state was antiferromagnetic, corresponding to the case depicted in Figure 2.9a. The initial and final states were different at temperatures of 120 and 130 K but only when the initial state was obtained by heating from 5 K. The initial and final states were the same at these two temperatures when the initial state was obtained by cooling from 300 K. The large isothermal variation required in p corresponded to H varying up to 8 T in this case.

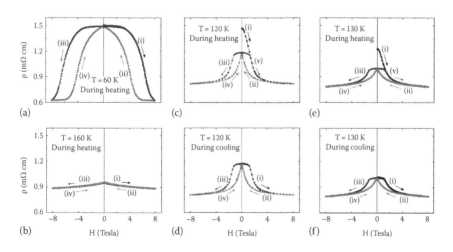

FIGURE 2.10

We show the data on $Mn_{1.8}Co_{0.2}Sb$ that corresponds to case (a). The open hysteresis loop is seen on the isothermal variation of H at T = 120 K (plots (c) and (d)), and at T = 130 K (plots (e) and (f)), when these T are reached by heating from 5 K but not when these T are reached by cooling. Open hysteresis loops are not observed at 60 K (plot (a)), which is below $T^*(0)$, and also not at 160 K (plot (b)), which is above $T^{**}(0)$. (Taken from Kushwaha, P. et al., *J. Phys. Cond. Matter.*, 20, 022204, 2008.)

The earlier discussion invoking two control variables highlights that the state of the system could be either metastable or stable depending on the history followed in the variation of control variables, and this reflects in terms of the value of q, which becomes path dependent. The free energy curves $G(q, p, T)$, however, are path independent. In this sense $G(q, p, T)$ are determined by thermodynamics as the possible allowed states, even though only the absolute minima in these correspond to equilibrium states. Since local minima in G occur in the control variable space bounded by the lines $p_{21}(T^*)$ and $p_{12}(T^{**})$, one would expect the experimental limits of supercooling and superheating to be independent of the control parameter that is being varied. This is often not so because we have a new feature emerging if $E_0(p)$ has a strong p-dependence. If we have a material in phase-2 well above p_{21} and are lowering p with a consequent sharp change in $E_0(p)$, then some fraction of this changing energy will be dissipated to the surrounding bath and some will be dissipated inside the material and contribute to the fluctuation energy E_f. This fraction will be small when there are no defects as for pressure variations in an ideal gas, and it will be large if variations in the magnetic field are moving vortices across pinning sites in a superconductor. Isothermal variations of p can thus restrict the extent of supercooling or superheating, causing the phase transition to be caused before p_{21} or p_{12} are reached [1].

Finally, this also causes oscillations in p to add to the fluctuation energy E_f, and we can now understand that applying such oscillations can cause a metastable state to transform to the stable state. The fluctuation energy would increase with increase in the amplitude of the oscillating ac field and also increase with increase in the duration for which the ac field was applied [1]. Varying any of these two (amplitude or duration) allows us to ascertain relative depths of the local minimum in $G(q)$ for different metastable states.

References

1. P. Chaddah and S.B. Roy, *Phys Rev* **B60** (1999) 11926; P. Chaddah and S.B. Roy, *Pramana J Phys* **54** (2000) 857. For experimental data, see S.B. Roy, P. Chaddah, and S. Chaudhary, *Phys Rev* **B62** (2000) 9191.
2. P.M. Chaikin and T.C. Lubensky, *Principles of Condensed Matter Physics*, Cambridge University Press, Cambridge (1995) Chapter 4.
3. T. Itami, J. Okada, Y. Watanabe, T. Ishikawa, and S. Yoda, *Mater Trans* **51** (2010) 1510.
4. J.D. Hunt and K.A. Jackson, *J Appl Phys* **37** (1966) 254.
5. X.S. Ling, S.R. Park, B.A. McClain, S.M. Choi, D.C. Dender, and J.W. Lynn, *Phys Rev Lett* **86** (2001) 712.
6. K. Huang and W. Eastern, *Statistical Mechanics*, Wiley Eastern, New Delhi, India (1975) p. 3.
7. P. Kushwaha, R. Rawat, and P. Chaddah, *J Phys Cond Matter* **20** (2008) 022204.

3

Defining Experimental Characteristics of First Order Transitions

3.1 Necessary and Sufficient Characteristics Following the Ehrenfest Classification

As discussed in Sections 1.2 and 1.3, the Ehrenfest classification gives prime importance to the lowest derivative of free energy that shows a discontinuity at the phase transition. A phase transition can be caused by varying two (or more) control variables, and the derivative of free energy corresponding to each of these control variables would show a discontinuity. For first order transitions, one should observe a discontinuity in the first derivatives of free energy, namely, volume, entropy, or magnetization. The discontinuities are related by the Clausius–Clapeyron relation.

Discontinuities in the first derivatives of free energy are sufficient to identify a first order phase transition and rule out a second order phase transition because the Ehrenfest classification requires the first derivatives to be continuous for a second order phase transition. The observation of a discontinuity in entropy at a (H, T) point (or a (T, P) point) requires a discontinuity in magnetization (or volume) at the same point.

The observation of a discontinuity in a property that is not a derivative of free energy, and thus is not an equilibrium property, does not identify the order of the transition. The most common example is a discontinuity in resistance, which is a nonequilibrium physical property. The superconducting transition can be either first or second order in different cases, but its most prominent signature is a sharp discontinuity in resistance, irrespective of the order of this transition. We concluded, however, in Chapter 2 that hysteresis is a signature of a first order transition. We shall pursue this in Section 3.3, but just stress here that *the modern classification allows us to treat all properties that are different in the two phases on the same footing*. This is not so in the Ehrenfest classification.

The investigation of vortex lattice melting provides a detailed look into how a first order transition was established experimentally, especially when there were conflicting theoretical models on the order of the transition.

This was pursued by various groups, following both the Ehrenfest and the modern classifications. Studies were being carried out on two high-T_C superconductors, $Bi_2Sr_2CaCu_2O_8$ and $YBa_2Cu_3O_7$ (referred to as BSCCO and YBCO, respectively). As discussed in Chapter 1, the latent heat and thus the entropy discontinuity was eventually measured at various points along the phase transition line in (H, T) space. Discontinuities were on the same line when scanning temperature and field. The discontinuity in magnetization was also measured at various points, and it was shown that the Clausius–Clapeyron relation was satisfied everywhere. This was thus confirmed as a first order phase transition at the highest level of rigor required by the Ehrenfest classification. However, various other measurements were taken prior to these ultimate measurements, and they indicated a first order transition and prompted researchers to perform more demanding and careful experiments. We shall discuss some of these earlier measurements.

3.1.1 Melting of the Vortex Lattice

The simplest measurement characteristic of superconductivity is that of zero resistance, but a small resistance (much lower than that of the normal state) is observed if the vortices of magnetic flux can move. This happens when the vortices get unpinned at current densities that are higher than the critical current density. This is also expected when the vortex lattice melts into a vortex liquid, as the vortices are then mobile, and there is no pinning at all even at the lowest current density. Among the first highly cited claims of a first order transition within the superconducting state of a high-T_C superconductor was the observation of a sharp change in resistance, ascribed to the freezing of the vortex liquid. This sharp change in resistance was hysteretic with both temperature and magnetic field as control variables.

Safar et al. [1] reported measurements of resistance as a function of both temperature and magnetic field on a defect-free single crystal of YBCO, using highly sensitive electronics with pico-volt sensitivity. When they applied a field of 6 Tesla parallel to the c-axis of the crystal, they observed a sharp change in resistance within the superconducting state. The resistance dropped around 81 K, from about 15% of its normal-state resistance to zero, over a temperature width of about 0.3 K. The sharp change in resistance was observed for both T and H as control variables, and in both cases there was a hysteresis between the increasing and decreasing scans for the control variable. The hysteresis was about 35 mK when the control variable was temperature and the magnetic field was fixed above 6 Tesla, and it was observed over the entire 300 mK over which this transition occurred. Similar results were seen in lower magnetic fields, with the temperature at which the transition occurred increasing and the hysteresis width decreasing. The hysteresis was not observed in the absence of a magnetic field, corresponding to a second order transition in that case. The broad transition was attributed to the vortex solid, consisting of a large number of "crystallites" of Abrikosov lattices that have melting temperatures spread over 300 mK due to inhomogeneities. It was conjectured by Safar et al. that all these crystallites were supercooling and superheating by the same amount, as this would explain the constant hysteresis width of about 35 mK [1]. Hysteresis, as the temperature is raised and lowered, is often observed as an artifact because the sample, the heater controlling the temperature, and the thermometer measuring the temperature are physically separated and this causes thermal lag (or backlash). Careful experiments are required whenever hysteresis is to be established, and differential thermocouples are often used to estimate thermal lags between the three points. Safar et al. claimed that the thermal lags in their system were significantly smaller than the hysteresis observed, and this was supported by the observed vanishing of hysteresis in 0 magnetic field [1]. They also showed a hysteresis of width 25 mTesla, with H as the control variable at T = 80.2 K. Finally, the extent of the hysteresis was reduced when the magnitude of the measuring current was increased. This was as expected because the measuring current was a disturbance, introducing fluctuations that enable crossing a larger free energy barrier.

This hysteresis in resistance, seen around those values of temperature and magnetic field where the resistance showed a sharp change, was thus reported as a strong indication of a first order transition, but this did not count in the Ehrenfest classification because resistance is not a thermodynamic equilibrium property and cannot be written as a derivative of free energy. Sharp changes in this physical property could not be used to determine the order of the transition in the Ehrenfest classification, which would require clear thermodynamic features like latent heat or a discontinuous jump in equilibrium magnetization.

Safar et al. [2] then followed up with measurements in higher values of H, going up till 16 Tesla as against 7 Tesla in their previous work. They reported that hysteresis persists till H = 10 Tesla, increasing as H is raised till 8 Tesla and then reducing to zero at 10 Tesla. This showed that the first order vortex lattice to liquid melting transition gives way to a second order vortex lattice to glass transition at H above 10 Tesla and that the first order melting transition is robust in the presence of a small amount of disorder. Again, the conclusions of the order of the phase transition were based on the observation of (or lack of) hysteresis, and not on the Ehrenfest classification.

Charalambous et al. [3] reported a hysteresis of about 7 mK in the resistance vs. T for a YBCO crystal. Increasing the transport current used to measure the resistance lowered the transition temperature in the heating cycle, indicating that this measuring current was adding fluctuations and the superheated state was transforming to the equilibrium state before reaching the limit of metastability. What was novel and interesting in this report was that unlike the temperature increasing cycle, the temperature reducing cycle showed a resistance independent of the measuring current, implying that this leg did not correspond to a metastable state. They could, thus, conclude that there was superheating but no supercooling in this vortex lattice melting transition—in striking contrast to the melting of a normal solid. This work, again, did not provide any evidence required by the Ehrenfest classification to support the conclusion of a first order transition. Similar resistive measurements were also reported in YBCO by Kwok et al. [4], with the hysteresis again reducing sharply at higher measurement currents. Hysteresis in resistivity continued to be used as the experimental signature for concluding that there was a likely first order transition.

Pastoriza et al. [5] then observed a discontinuous change in the magnetization with varying temperature in a single crystal of BSCCO, thereby providing evidence of discontinuity in a first derivative of free energy. It was probably the first experimental evidence consistent with the Ehrenfest classification, but they noted that the precision of the data was not enough to provide the T- and H-dependence of this discontinuity. More than 2 years had passed since the initial report by Safar et al. [1], with many experimental groups attempting to settle whether this transition within the superconducting state was first order. The report by Pastoriza et al. [5] was indeed a conventionally acceptable indication of a first order transition, but, as noted later by Zeldov et al. [6], "the observed change in magnetization was relatively broad and not well quantified." This was the first measurement reporting a discontinuity in an equilibrium physical property, and that too was relatively broad. The discontinuity in entropy was yet to be measured, and the Clausius–Clapeyron relation needed to be checked. As noted in Chapter 1, there were theoretical works predicting that the vortex melting transition should be second order. What kept the experimentalists active in challenging these theoretical observations were the credible observations of hysteresis in resistance—a nonequilibrium property.

As described briefly in Chapter 1, the issue was settled beyond doubt within the Ehrenfest classification. Zeldov et al. [6] noted that the local induction B inside the single crystal sample was not uniform, akin to that inside a solid experiencing nonuniform pressure across the sample. For a fixed, externally applied H, with the B experienced at various positions being different, the phase transition temperature would be different at different points. Implicit here is the concept that we are dealing with a first order transition that proceeds by large amplitude changes in order parameter but over short lengths, or by nucleation of the new phase over small correlation lengths. The concept of transition occurring in local regions in a sample is specific to first order transitions (and is not acceptable for second order transitions), and we shall be using it in good measure—it has been highlighted by us over the years [7–10].

The signal in a standard macroscopic measurement would average over the entire sample, and the observed signature of a phase transition could be broader and smoother than the underlying physical signature. Zeldov et al. [6], therefore, attempted local measurements of high spatial resolution; the limits in technology allowed them to probe the local magnetic field across a BSCCO crystal over small areas of about 3 μm × 3 μm using an array of microscopic Hall sensors. At a fixed temperature of 80 K, they scanned the applied field and observed steps in the local field measured by various sensors. Steps occurred in different sensors at different values of the applied field, but steps in each sensor occurred at the same value of the local field. This clearly brought out that the broad transition being observed was due to intrinsic inhomogeneous local fields, and not due to some other artifact. They also kept the applied field fixed and scanned the temperature as they measured the local field in each sensor. Again, steps were observed in different sensors at different temperatures, but the values of the local field and the (global) temperature at which steps were observed were falling on a (H_C, T_C) curve. The steps in equilibrium magnetization confirmed the first order nature of the transition, and the observation over finite lengths confirmed nucleation with finite correlation length, as expected for such transitions.

Schilling et al. [11] subsequently reported macroscopic measurements on a single crystal of YBCO, where they measured the jump in equilibrium M as H was varied isothermally. This was not a sharp discontinuity but displayed the behavior shown schematically in Figure 3.1, with a region in H over which the magnetization changes sharply. In the macroscopic measurements by Schilling et al. [11], this change was shown as occurring over a field range of 0.1 Tesla at an H around 3.7 Tesla. Measurements by Welp et al. [12], again on a YBCO crystal, also showed a similar width of about 0.1 Tesla for the "discontinuous jump" in magnetization. They also reported that for iso-field measurements with temperature being varied, sharp changes in magnetization and resistance occurred at the same (H,T) points as in isothermal measurements. We should emphasize that the local

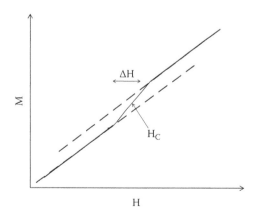

FIGURE 3.1

We show a schematic for isothermal M-H, as reported in References 11 and 12, for a single crystal of $YBa_2Cu_3O_{7-d}$. Isothermal scans were reported at temperatures ranging from 82 to 90 K. The "jump" in magnetization occurs over a ΔH of about 0.1 Tesla for T = 84 K. This is over 2% of the value of H_C, which was about 4.2 Tesla. The micro-Hall array probe used by Zeldov et al. on BSCCO showed a visually sharper jump with a relative width ($\Delta H / H_C$) lower by one order of magnitude.

measurements of Zeldov et al. [6], albeit in a different material, showed changes in magnetization over a smaller range of 0.1 Oe at a field of 58 Oe.

Schilling et al. [11] also made macroscopic measurements of the latent heat across this transition, both with T and with H as the control variable. They measured the temperature on the sample and on a reference material as the control variables were varied. When the sample made the transition from the higher entropy to the lower entropy phase, latent heat was released, the sample temperature became higher than that of the reference material, and they equilibrated to the same value after the transition point was crossed. This should happen both as T is lowered and as H is reduced. When the sample makes the transition from the lower entropy to the higher entropy phase, as T is raised or as H is increased, latent heat has to be absorbed and the sample temperature decreased to lower than that of the reference material. This latent heat was not absorbed (or released) at a fixed value of the control variable H, as should happen for a first order transition. It was absorbed (or released) over a range of about 0.1 Tesla when the transition was occurring at 4.9 Tesla, and the integrated value was used as a measure of latent heat. We consistently see that this first order melting transition has a width, reinforcing the statement of Zeldov et al. [6] that "the observed change in magnetization was relatively broad." We shall look again at the conjecture of Safar et al. [1] that the broad transition was due to the vortex solid consisting of a large number of "crystallites" of Abrikosov lattices that have melting temperatures that are spread because of inhomogeneities.

Schilling et al. [11] could measure the latent heat at this transition calorimetrically, by traversing through the transition, varying the applied field isothermally as well as by varying T at fixed values of H. The two types of scans gave consistent values of entropy jump (or latent heat) along the (H_C, T_C) phase transition curve. These provided exhaustive measurements of latent heat for the vortex lattice melting transition, and the simultaneous measurement of magnetization jumps on the same sample enabled a check that the Clausius–Clapeyron relation was satisfied [11].

3.2 Need for Other Characteristic Tests beyond the Clausius–Clapeyron Relation

These studies on the nature of the phase transition accompanying the melting of the two-dimensional vortex lattice in high temperature superconductors highlight that a hysteresis in the discontinuous jumps in resistivity was accepted as the indicator of a first order transition, even though theoretically it was still under discussion whether this should be a first order or a second order transition. Since resistivity is not a thermodynamic property, this could not be considered evidence in the Ehrenfest classification. Discontinuities in equilibrium magnetization and entropy, with latent heat satisfying the Clausius–Clapeyron relation at various points on the phase transition line, are the accepted confirmatory evidence for a first order transition consistent with the Ehrenfest classification. This was provided for the vortex lattice melting transition only about 4 years after the first validated reports on hysteresis at the resistivity discontinuity. The resistivity measurements showing hysteresis, including its dependence on the measuring current, provided results that kept motivating more experiments to confirm a first order transition.

As was brought out in earlier text, measuring a discontinuity in an equilibrium property can be a daunting experiment. As pointed earlier, it has been recognized that there are experimental difficulties when the latent heat is small and it becomes difficult even to distinguish it from a peak in specific heat [13]. There is also an experimental problem in measuring a discontinuity when the jump in magnetization is small, and more so when the transition is broadened by disorder and the discontinuity becomes only a sharp change. Measurement of magnetization discontinuity in BSSCO had to overcome the fact that different regions of the sample do not show the transition at the same value of the applied field. The work of Zeldov et al. [6] thus showed that the jump or discontinuity reported in the macroscopic measurement by Pastoriza et al. [5] was only a sharp change. Some years later, Soibel et al. [14] showed in BSSCO crystals that there is actually a landscape $H_m(x, y)$ and $T_m(x, y)$ at which the local transition would be seen. The spatial distribution of (H_m, T_m) across the sample was attributable to a quenched disorder, consistent with what was proposed by Safar et al. [1]. Disorder-broadened first order phase transitions have many new features, which are discussed in the subsequent chapters. Jumps and discontinuities in the derivatives of free energy are not observed, but the broadened transition will show hysteresis in physical properties that may not even be derivatives of free energy.

As has been noted earlier in Reference 15, most of the functional magnetic materials of interest are multicomponent systems whose properties become more interesting with substitutions, and thus these have intrinsic frozen disorder. This has resulted in studies on many materials that show "broad first order transitions," a phrase that would be considered an oxymoron, or self-contradictory, in the Ehrenfest scheme. Even though the modern classification is considered to owe its origin to transitions that cannot be identified as of any integer order, in this book we are going beyond the Ehrenfest classification as a result of experiments on broad first order transitions that have almost become ubiquitous. In the case of a material having frozen disorder, there would be a landscape of free energy densities. The sharp (T_C, H_C) boundaries would now be defined only over individual regions having length scales of the order of the correlation length. This would result in a rounding of the discontinuity when macroscopic measurements of physical properties are made. A very thorough observation of such a behavior was reported by Roy et al. [16] in a magnetic transition in a sample of doped $CeRu_2$, where macroscopic and mesoscopic measurements using a micro-Hall probe were made with both H and T as control parameters. The transition clearly progressed over a range of values for both the control variables, establishing the existence of landscape $H_m(x, y)$ and $T_m(x, y)$ values at which the local transition would be seen. Hysteresis was also reported in all these measurements, but, as stressed earlier, hysteresis measurements have no significance in the Ehrenfest classification. The studies by Roy et al. do not enable calling this a first order transition in the Ehrenfest classification. The significance of hysteresis—and its dependence on measurement protocols—will be discussed in what follows, in the context of the modern classification of phase transitions.

3.3 Necessary and Sufficient Characteristics of a First Order Transition following the Modern Classification

3.3.1 Hysteresis as an Indicator

In the modern classification, we start with the two distinct phases that exist on either side of the phase transition having different order parameters. As one approaches the phase transition by varying a control variable toward its critical value, the distinction between first order phase transitions and continuous phase transitions is seen. In a continuous phase transition, the order parameters merge and all physical properties become identical at the phase transition point. At this point, the two phases are identical, with no memory of how the transition point was reached, i.e., from phase-1 or from phase-2. On crossing the phase transition point, this lack of memory ensures that there can be no supercooling or superheating. Any indication of supercooling or superheating that is not ascribable to an experimental artifact, such as a thermal lag, implies that the transition is not continuous. A continuous phase transition cannot show hysteresis in any property. Hysteresis in a physical property across the value of a control variable at which a sharp change in physical properties is observed is thus an indication of a first order transition and rules out a continuous phase transition.

We caution here that hysteresis by itself can be observed without an underlying phase transition because of an intrinsic lag—as distinct from a lag caused by the experimental artifacts mentioned earlier—in the physical response of a material. This hysteresis does not coincide with any sharp change in physical properties. The most common example is hysteresis seen in the M-H (magnetization vs. field) response of a ferromagnet (FM), which is attributed to the pinning of domain walls and the consequent hindrance to the realignment of domains. The strength of this domain wall pinning is reflected in the coercive field H_C, as depicted in the schematic Figure 3.2. Hysteresis changes with changes in the temperature at which the isothermal M-H scan is measured, as H_C reduces with rising T. The relevant physical

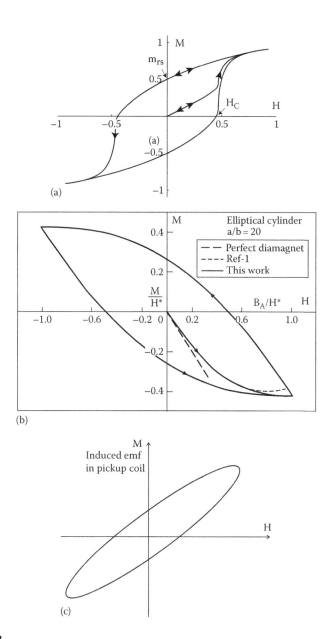

FIGURE 3.2
We show three examples where hysteresis is not due to an experimental artifact like thermal lag but is a manifestation of hindered kinetics. In (a), we show the schematic M-H curve of a ferromagnet, depicting hysteresis caused by pinning of domain walls. In (b), we show the calculated M-H curve of a superconductor with a finite value of J_C. This hysteresis is consistent with the Critical State Model and is due to the pinning of vortices. In (c), we show the schematic M-H curve of a conductor put in an ac magnetic field with a suitably high frequency, as measured using a pickup coil. This hysteresis is due to the "skin effect." (b: Taken from Bhagwat, K.V. and Chaddah, P., *Phys. Rev.*, B44, 6950, 1991 [17].)

property is magnetic susceptibility, and there is a continuous phase transition at the temperature where there is a sharp change in magnetic susceptibility. There is no hysteresis in a thermal scan because the phase transition is continuous and is not first order. Hysteresis in the isothermal scan of magnetic field is attributed to the FM domains switching the direction of their polarization to align with the direction of the applied field. The hysteresis in the M-H curve is due to the transformation from a FM state with spin down to that with spin up and not due to a phase transition.

A more specialized but extensively studied example is the hysteresis in M-H curves within the superconducting state. This has been well understood by using Bean's critical state model, which allows the pinning of vortices until the gradient in their density corresponds to the critical current density J_C in the superconductor. Here, the kinetics of vortices is hindered by critical pinning force $F_P = J_C \times B$, and there would be no hysteresis if the pinning force were 0. The third example is the response of any metal to an alternating magnetic field H_{ac}, where the signal M_{ac} induced in a pickup coil surrounding the sample shows a phase lag. A plot of M_{ac} vs. H_{ac} is an ellipse, and the area enclosed within this ellipse corresponds to a hysteretic loss. This area shows a maximum as the frequency of H_{ac} is varied, and it corresponds to the skin-depth of the AC field scanning across the dimensions of the sample. There is no sharp change in any physical property and no underlying phase transition, and the hysteresis can again be understood as due to a phase lag associated with the velocity of light being finite. In some sense, this can be attributed to the propagation of electromagnetic waves being hindered. To sum up, hysteresis across a phase transition does imply a first order transition, but hysteresis is obviously also possible due to hindered kinetics, intrinsic to the material, causing a delayed response.

By itself, the observation of hysteresis cannot be used to assert that a first order transition exists. The existence of a metastable state, and a different stable state, at the same point in the control variable space is the essential feature of hysteresis associated with a first order phase transition. The emphasis will be on establishing that the state of the system depends on the route (or path) followed in the control variable space, because both the metastable and the stable states can exist at the same point in this space. It will be necessary to establish that only one of these path-dependent states is stable, or to use the results obtained in Chapter 2 to show that the path-dependence (or route dependence in the 2-control variable space) is consistent with an underlying first order transition.

3.3.2 Path-Dependent Metastable and Stable States

Let us first consider an example where sharp changes in physical properties, along with observation of hysteresis, have been used to identify a transition as first order. It should be stressed that in the context of first order magnetic transitions in magnetic materials—as distinct from first order magnetic transitions in superconducting materials discussed earlier—the stringent requirement of measuring latent heat and checking for compliance with the Clausius–Clapeyron relation has neither been sought nor tested experimentally in recent years. This is true for many magnetic materials that have been investigated in the last two decades or so. The contrast is surprising but can be understood as the research here was driven mostly by potential applications and not by basic concepts as in the case of vortex lattice melting. This may not be true for the case of half-doped manganites that we now discuss. Here the research was driven by the colossal magnetoresistance these materials displayed and their possible application as magnetic memory materials. The transition, however, was attributed to the new concept of "melting of charge-order." The research was then dominated by observations of coexisting "charge-ordered" and "charge-liquid" phases at very low temperatures, well below the temperature regime in which the transformation was initiated during cooling. Explanations offered included the possibility of an inhomogeneous ground state to obviate the problem of nonzero entropy at 0 K.

We shall first describe the limited observations that were made to assert a first order transition in the title of the paper [18] that was, surprisingly, accepted and published in the period intervening between the publication of Zeldov et al. [6] and Schilling et al. [11], i.e., when the need for measuring latent heat and satisfying Clausius–Clapeyron relation before establishing a first order transition in vortex lattice melting was being so prominently addressed.*

Kuwahara et al. [18] were studying a transition in a half-doped manganite material, $Nd_{0.5}Sr_{0.5}MnO_3$ (NSMO), which represented the so-called "colossal magnetoresistance (CMR) manganites." They found a transition at T = 158 K in which resistance increased by orders of magnitude during cooling, with a simultaneous disappearance of ferromagnetism. The changes in both these physical properties were sharp and occurred at 158 K. In addition, they observed a sharp change in the lattice parameters at the same temperature. They did not report observation of hysteresis in any of these three properties in temperature scans. The higher entropy phase-2 was a FM metal, and the lower entropy phase-1 was an antiferromagnetic (AFM) insulator. They studied the sharp change in resistance—as they varied the magnetic field isothermally—at various temperatures below 158 K. At low values of H the resistance was large, corresponding to an insulating state, and it dropped by several orders of magnitude when H was raised above a critical value, H_{12}, showing a transition to a metallic state. On lowering the value of H, the resistance increased sharply at a critical value, H_{21}, showing the reverse transition to the insulating state. At all temperatures where this isothermal scan was made, they obtained $H_{21} < H_{12}$ as expected for hysteresis across a first order transition. This was ascribed to a

---

* Reference [18] was accepted on 30 August 1995 and was published on 10 November 1995, in the period intervening between the publication of reference [6] on 01 June 1995 and reference [11] on 29 August 1996.

first order transition. We caution that hysteresis in magnetization during such isothermal scans in hard ferromagnets and in hard superconductors has just been described earlier without an underlying first order transition.

We now describe an anomaly in the data shown in Reference 18 that led us to the development of the ideas in the following chapters of this book. At temperatures below 158 K, both H_{21} and H_{12} increased with reducing temperature. H_{12} increased from about 2.5 Tesla at 140 K to about 8 Tesla at 80 K, while H_{21} increased from about 1.5 Tesla to about 6 Tesla. The isothermal transitions were sharp and hysteretic, and this clear hysteresis was accepted by the authors as evidence for a first order transition. We, however, noticed some anomalies at lower temperatures. At 60 K, the corresponding values were about 8.5 Tesla and about 6 Tesla. At lower temperatures, H_{12} kept increasing, rising to over 11 Tesla at 2.5 K, while H_{21} started anomalously decreasing, falling to below 2 Tesla at this temperature. This is depicted in the schematic in Figure 3.3. This sharp reduction in H_{21} while H_{12} was increasing was explained by Kuwahara et al. as being caused by reducing thermal fluctuations [18], but we found the explanation to be convoluted.

Further, resistivity values obtained at such low temperatures and high fields were noted by the authors as "remarkably route dependent" [18]. The resistivity corresponded to an insulator when cooled to 5 K in a constant H = 4 Tesla, but it corresponded to a metal when measured after cooling to 5 K in H = 7.5 Tesla and then lowering H to 4 Tesla in an isothermal scan. A proper explanation for this remarkable route dependence will be brought out in the subsequent chapters. In a later section, we discuss more detailed measurements that scanned both H and T.

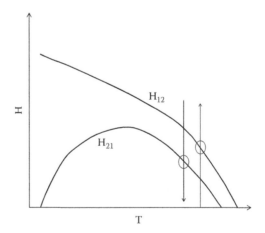

FIGURE 3.3
We show a schematic of the isothermal data reported by Kuwahara et al. [18]. H_{12} line corresponds to the sharp drop in resistance for the insulator-to-metal transition with an isothermal increase in H, while the H_{21} line corresponds to the hysteretic reverse transition with an isothermal decrease in H. The transition points are indicated by circles.

The transition in NSMO was from an FM metal to an AFM insulator and was caused because doping created an equal number of Mn^{3+} and Mn^{4+} ions. With lowering of temperature, the location of these ions changed from a random occupation of specific sites to an alternate ordering. The transition from an FM metal to an AFM insulator is attributed to the ordering of these charged ions and the reverse transition to the "melting" of this charge-ordered lattice. As expected for transitions from one magnetically ordered state to another, H was a control variable and increasing H would cause a transition from an AFM to an FM state and cause the melting of charge order. Kuwahara et al. [18] studied the transition points as H is increased and decreased by resistivity measurements only, and they reported the observed hysteresis in resistivity as evidence that the transition was first order. We remind ourselves that resistivity is not a thermodynamic equilibrium property and emphasize that hysteresis also needs to be observed when the transition is monitored by the change in magnetization with varying H. Further, one must also observe thermal hysteresis, or hysteresis when T is the control variable, while measuring both resistance and magnetization. It is also necessary to show that the route dependence of a physical property in the hysteretic region is consistent with predictions of an underlying first order transition, based on which state is metastable. The acceptance of the conclusion of a first order transition in Reference 18 without these additional checks is a clear indication that the criteria for establishing a first order transition were still evolving in that period.

A proper explanation for the route dependence that Kuwahara et al. [18] observed will be brought out in a later section based on more detailed measurements that scanned both H and T, in the same material (NSMO) and in a suitably selected material (Co-doped Mn_2Sb [20]), where the transition starts around 140 K and H_{21} started decreasing below 50 K. As we shall see in subsequent chapters, hysteresis and route-dependence that was observed in these materials was actually a complicated admixture of hysteresis caused by a first order transition and that caused by hindered kinetics. We explained the drop in H_{21} below a certain temperature as due to kinetic arrest and subsequent de-arrest. This is further discussed in Chapter 4.

3.3.3 Hysteresis Together with Metastable to Stable Transformation

In a first order transition, the order parameters of the two phases remain distinct even at the phase transition point. Two phases with distinct order parameters, with distinct first derivatives of free energy, and with some other physical properties also being different, can coexist at the phase transition point. As described at the end of Chapter 2, hysteresis is observed because the phase that exists at and in the near proximity of the transition point depends on the history of how that point was reached. This implies that hysteresis would be observed in all the physical properties that are distinct in the two phases.

This hysteresis occurs because at some values of the control variables, and under some suitable path (or thermo-magnetic history) of reaching those values, we observe supercooling or superheating and the material is in a metastable state. The understanding of a first order transition described earlier dictates that under some other history the material goes into the stable thermodynamic equilibrium state. This also implies that in the hysteretic region of the (H, T) space, the system can be put into two distinct states, one of which is metastable. The schematic in Figure 3.4 describes paths by which we can have the system in the metastable state corresponding to the

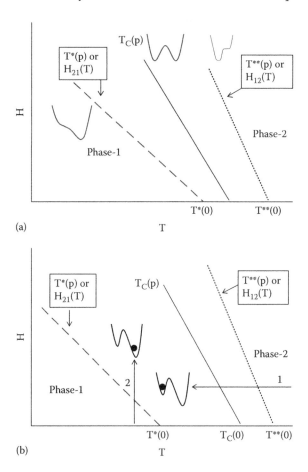

FIGURE 3.4
We show in (a) a schematic for a first order transition in vortex matter, where the transition temperature drops as the magnetic field rises. A linear dependence is assumed for simplicity and without loss of generality. The limit of supercooling is shown by the $H_{21}(T)$ line, and the limit of superheating is shown by the $H_{12}(T)$ line. Free energy curves corresponding to these and to the phase transition line are depicted. Cooling in fixed H, called the field-cooled (FC) state, is depicted in (b) by path-1 and results in a metastable phase-2. The occupied state is depicted by the filled circle. Cooling in H = 0 and then raising H isothermally, called the zero-field-cooled (ZFC) state, is depicted by path-2 and results in the stable state (phase-1). The metastable and stable states are depicted by the filled circles in the local and the absolute minima in the free energy curves.

local minimum in free energy or in the stable state corresponding to the global minimum in free energy. Following the discussion toward the end of Chapter 2, the metastable state at a particular (H, T) point can be converted into the stable state at the same (H, T) point by introducing fluctuations. Small fluctuations introduced in a stable state do not convert it into a metastable state. (A stable-to-metastable conversion requires unidirectional variation of large amplitude in the control variable causing the phase transition and then a reverse variation back into the supercooled or superheated state.) The metastable state becomes unstable under fluctuations, and these can be introduced through pressure oscillations in a solid, stirring in a liquid, or variations in the magnetic field for a magnetic material.

The observation of two distinct physical properties at the same value of the control variables (H, T), with one of these being unstable and the other stable under the same fluctuations caused by variations of the magnetic field, is thus a clear indicator of the magnetic transition being first order.

We have described in the previous section how hysteresis in resistance was pursued as an indicator of a first order transition. In early experiments, Chaudhary et al. [20] observed, within the superconducting state in single crystal samples of both $CeRu_2$ and V_3Si, zero resistance in the field-cooled (FC) mode but finite resistance in the zero-field-cooled (ZFC) mode. This multivalued path-dependent resistance was attributed to a first order transition between two states of a vortex solid with different pinning properties and, thus, different values of critical current density. Critical current density J_C, like resistance, is not a thermodynamic equilibrium property. They also argued that by following two different paths in the (H, T) space, they were reaching states with two different values of J_C over some region of H, and one of these (H, T) paths had resulted in a metastable state. This should have converted to the stable state, which resulted from the other (H, T) path, if sufficient fluctuations could be introduced to overcome the free energy barrier. Chaudhary et al. [20] showed that cycling the magnetic field though 25 mTesla, with the background field being over 1 Tesla, was enough to change the FC state to the ZFC state in $CeRu_2$. For the V_3Si crystal, the background field was over 2 Tesla, and a pulse of 50 mTesla was enough to change the FC state to the ZFC state. Thus, the FC path was resulting in a supercooled higher entropy phase that was metastable, while the ZFC path resulted in the lower entropy stable phase, as depicted in the schematic Figure 3.4.

The region of (H, T) over which the two thermomagnetic history–dependent states could be observed was also studied by Roy et al. [21] in the same crystal of $CeRu_2$. They found that the isothermal variation of H to cross the phase transition resulted in a smaller (H, T) region where this metastable supercooled state was seen, compared to the supercooled region under constant-H variation of T to cross the phase transition. This was consistent with the behavior expected of metastability across a first order transition between two states of vortex solid. Variation of H moves the vortices, adding to fluctuations and making the metastable state unstable in the region where the free energy barrier is small.

We now describe a very thorough experiment on the vortex lattice phase transition that succinctly brings out these ideas that had been pursued in all the earlier works and establishes that vortex melting is a first order phase transition. In the modern classification of phase transitions, this experiment was as complete for establishing a first order phase transition as the work of Schilling et al. [11] was in the Ehrenfest classification.

Ling et al. [22] did a small angle neutron scattering (SANS) study of the superconducting state in a niobium single crystal. Note that this is a conventional superconductor with large coherence volume, and fluctuation effects would be small. Melting of the vortex lattice would thus be more subtle than that for the high-T_C superconductors. The SANS measurement provides a diffraction pattern that reflects the structure of the vortex lattice. The diffraction pattern for a vortex liquid is a circular ring, as depicted in Figure 3.5a, and that for a vortex lattice is a hexagonal array, as depicted in Figure 3.5b. The (H, T) values at which a phase transition occurs between these two were estimated by ac-susceptibility measurements, and the onset during heating of the sharp minimum in $\chi(T)$, measured in constant H, was taken to be the first order vortex lattice melting transition. This was then subjected to detailed tests that checked for states being metastable. At a constant field of H = 3.75 kOe the transition under varying temperature was at about T = 4.5 K. Interpolating such data, they estimated that if the temperature was kept constant at T = 4.4 K, the transition would occur at a field of about H = 3.9 kOe. With this data from ac-susceptibility measurements, Ling et al. performed some very innovative measurements. First, they measured the SANS pattern at H = 3.75 kOe and T = 3.5 K, which showed that the vortex structure was of a hexagonal lattice. They also measured the SANS pattern at H = 3.75 kOe and T = 4.6 K, which showed a ring corresponding to a vortex liquid. Both these patterns were taken by cooling in fixed H, and the schematic Figure 3.5 indicates the pattern for these cases and the important conclusion that T = 3.5 K lies below, and T = 4.6 K lies above, the supercooling limit T^* for H = 3.75 kOe. They surmised from their ac-susceptibility data that even T = 4.4 K was below the melting point T_m = 4.5 K but above T^* for this H, and that at H = 4.00 kOe, T = 4.4 K was above T_m = 4.30 K but below the superheating limit T^{**}. They then checked this surmise as described in the following text.

As indicated in Figure 3.4, these two points can be reached by various paths. If their surmise about the location of these points was correct, then the material could be either in the metastable state or in the stable state, depending on the path followed varying the control variables H and T. They chose two paths, namely, cooling to the appropriate temperature with H = 0 and then increasing H to the appropriate value isothermally (ZFC path), and applying the appropriate H at high T and then cooling to the appropriate T in constant field (FC path). At both points, the ZFC path showed spots corresponding to

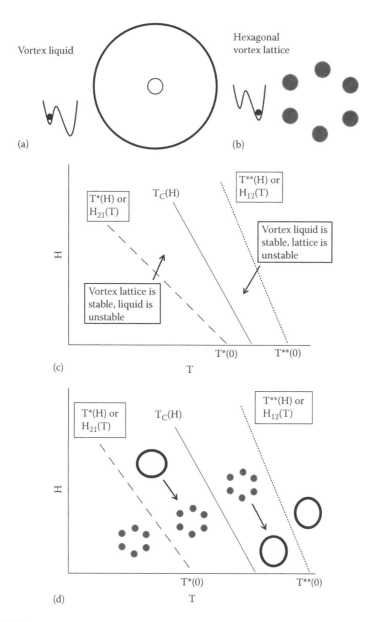

FIGURE 3.5

We show in (a) the schematic diffraction pattern from the two-dimensional vortex liquid, which is the higher entropy phase and has a zero-order parameter, and in (b) the schematic diffraction pattern from the two-dimensional hexagonal vortex lattice, which is the lower entropy phase and has a finite-order parameter. In (c), we depict the regions of (H, T) where the observed state is path-dependent, and in (d) we depict the metastable to stable transformation that can be observed by introducing fluctuation through variations in the magnetic field. The reverse transformation cannot be observed in the respective regions. Only the hexagonal lattice exists below the T* line, and only the liquid exists above the T** line.

a hexagonal lattice of vortices while the FC path showed these smudged on a ring corresponding to a liquid of vortices or to a disordered structure with short-range order. Consistent with the melting at T_m being a first order transition, one of the two states must be metastable at each of these (H, T) points.

We first consider H = 4.00 kOe and T = 4.4 K and check if this was above T_m = 4.30 K but below the superheating limit T**. If so, then the ZFC path should give a superheated state and the vortex lattice observed should be metastable while the vortex liquid observed by following the FC path should be stable. Ling et al. [23] applied an ac field of amplitude H_{ac} = 3.3 Oe and frequency v = 100 Hz to both these states. The total energy fluctuation introduced into the material is proportional to $H_{ac}^2 v \tau$, where τ is the time for which the ac field of frequency v is applied. By in-situ SANS, they found that the peaks in the pattern start to disappear at τ = 10^2 seconds, and disorder sets in and the pattern becomes the same as that obtained by FC path after $\tau = 10^3$ seconds. This is consistent with the ZFC path giving a metastable state and the FC path having given the stable state. To confirm that the FC path had given a stable state, the same H_{ac} was applied for $\tau = 10^4$ seconds on the state obtained in the FC path and no change was observed.

Now consider H = 3.75 kOe and T = 4.4 K and check if this was below T_m = 4.50 K but above the supercooling limit T*. If so, then the FC path should give a supercooled state and the vortex liquid observed should be metastable while the vortex solid observed by following the ZFC path should be stable. The free energy barrier surrounding the supercooled state was higher than that surrounding the superheated state. Ling et al. [22] used the superconducting magnet itself to apply a large amplitude field cycling of H_{ac} = 50 Oe and frequency v = 0.1 Hz to both these states. The ZFC state was unaffected, but the FC state showed a very dramatic appearance of spots with six-fold symmetry. The metastable vortex liquid had converted to a vortex lattice under this field cycling. The conversion of the metastable FC state to the stable ZFC state by cycling the magnetic field was reminiscent of the work of Chaudhary et al. [20] on $CeRu_2$ described earlier, as was the explanation and motivation, but the data of Ling et al. [23] was visually striking and very convincing. With this data, measuring latent heat or satisfying Clausius–Clapeyron relation was almost superfluous. The work of Ling et al. [22] establishes metastable to stable transformations, and presents a complete test for classifying a phase transition as first order.

Thus, going beyond just observing hysteresis, one should be able to identify a region in the control variable (H, T) space where two different states can be reached by following different paths in this space. One has to identify physical properties that are drastically different in these two states. Then, one needs to show that one of these states is metastable under the application of disturbances that add fluctuations, and that the physical property identified earlier will change to the value corresponding to that in the stable state. This provides a convincing claim of a first order transition when latent heat cannot be measured either because it is small or because, as we have mentioned in Section 3.2 and shall discuss in detail in Chapter 5, the first order transition occurs over a range of values of the control variables.

Following the procedures mentioned earlier convincingly identifies a transition as first order but, as noted repeatedly earlier, observation of hysteresis (that is not an experimental artifact) is an indicator that a transition is first order. However, the absence of hysteresis or of metastability is not sufficient to assert that a transition is not first order or is a continuous transition: hysteresis will be absent if the fluctuation energy E_f is large enough to prevent the existence of metastable states, as when a stirrer is vibrating in a liquid being heated to boiling.

Comparatively exhaustive measurements of hysteresis that fall between the completeness of Ling et al. [22] and the indicative results of Kuwahara et al. [18] are now performed frequently before identifying a magnetic transition as first order. Hysteresis is routinely reported both in resistance and in magnetization, even though a magnetic transition is so characterized because of a sharp change in magnetization. This is because an FM to AFM transition doubles the unit cell, changes the Brillouin zone, and consequently increases the resistance. Thus, sharp changes in magnetization in metallic systems are usually also accompanied by sharp changes in resistance.

We consider some examples of hysteresis measurements that go beyond the limited measurements of Kuwahara et al. [18]. Studies on doped $CeFe_2$ (with Al and with Ru doped on Fe site) were the first ones that reported hystereses in both resistivity and magnetization and established that the data of Kuwahara et al. was actually a complicated admixture of hysteresis caused by a first order transition and hysteresis caused by hindered kinetics. In a series of papers on a $CeFe_2$ sample with Al-doping, Manekar and coworkers [7,23,24] observed anomalous behavior in isothermal M-H and R-H measurements, where the initial virgin curve was observed to lie outside the envelope hysteresis loop. The anomalies were observed below 20 K and became more pronounced as T was lowered to 2 K. They also observed anomalous behaviors in the temperature-dependent measurements of both magnetization and resistance with H held constant at various values. They ascribed these anomalies to the existence of metastable supercooled and superheated phases across the transition boundary but complicated by the kinetics of the transition being hindered and even arrested at lower temperatures. We shall address such behavior in Chapter 4. They argued that some of the historic effects observed across the FM to AFM transition by Kuwahara et al. [18] could be explained by this proposed interplay of supercooling and hindered (or arrested) kinetics and tested their explanation by following an unusual history dependence of the external magnetic field. These were then followed by studies on a Ru-doped sample of $CeFe_2$ [25,26] to confirm the phenomenological model. These will be discussed in Chapter 5 after introducing new concepts.

As discussed in Section 2.6, Kushwaha et al. [19] studied a magnetic first order transition in Co-doped Mn_2Sb, where the lower entropy state was AFM and where the transition started around 140 K. As in the observations of Kuwahara et al. [18] on NSMO, they found that both H_{21} and H_{12} increased as T was lowered below 140 K, but H_{21} started decreasing below 50 K. The resistance in the initial state at H = 0 and the remnant state after an isothermal scan to 8 Tesla were different at 130 K as the remnant state was metastable. The difference increased as the temperature was lowered from 130 to 120 K but started reducing as the temperature was reduced, and the difference was not measurable between 100 and 30 K. The difference was again measurable at T = 30 K, and increased monotonically as T was lowered to 5 K. They showed that the path-dependence around 120 K was due to supercooling while that below 30 K was due to arrested kinetics. We shall come back to how this was concluded in subsequent chapters.

Rawat et al. [27] also studied polycrystalline samples of the same material (NSMO) that Kuwahara et al. [18] had studied in its single-crystal form. They showed sharp changes in T-dependent resistance and magnetization in various magnetic fields. They showed hysteresis between cooling and heating cycles in resistance for H ranging from 0 to 2 Tesla and in magnetization for H ranging from 0.05 to 2 Tesla. It is to be noted that magnetization requires a bias field for measurement while resistance can be measured in zero bias field. The temperatures for the sharp changes were the same for both M and R, in H = 1 Tesla and also in H = 2 Tesla. They also showed anomalies in isothermal M-H and R-H curves with the initial virgin curves lying outside the envelope hysteresis loops. The anomalies became less pronounced as T was raised from 5 to 30 K and disappeared at 100 K in both cases. They also showed anomalies in the T-dependence of the magnetic field at which the return transition takes place in isothermal M-H measurements. While reproducing results of Kuwahara et al. [18], they showed further features supporting the existence of metastable supercooled and superheated phases across the transition boundary, together with the kinetics of the transition being arrested at lower temperatures. These features, and their origin, will be discussed in detail in the subsequent chapters.

The association of hysteresis with a first order magnetic transition has become common, and we have come across studies where the absence of hysteresis in a magnetic transition is used to classify the transition as a continuous transition. As has been emphasized earlier, absence of observable hysteresis does not imply that a transition is not first order. A transition cannot be characterized as a continuous transition just because some physical property that changes sharply at the transition does not exhibit hysteresis.

To sum up, since hysteresis is not possible in any physical property across a continuous phase transition but can occur in various physical properties across a first order phase transition, observation of hysteresis in the measurement of a physical property that changes sharply does indicate a first order transition. If this hysteresis is not an experimental artifact, then it must not be seen at values of the control parameter away from the transition point. Hysteresis should also not be seen at the existing values of the control parameter after using the second control variable to shift the transition to a different value of the first control parameter. These are some essential experimental checks that must be performed. Hysteresis is, however, also observed because of arrest of kinetics, as was discussed in the case of pinning of FM domains, pinning of vortices in superconductors at current densities below J_C, and because of skin effect in conductors exposed to an AC field. *We must not only ensure that the hysteresis is not due to experimental artifacts but also confirm the cause of the observed hysteresis.* Hysteresis shown to be associated with the existence of supercooled or superheated states is sufficient to establish that there is a first order phase transition.

References

1. H. Safar, P.L. Gammel, D.A. Huse, D.J. Bishop, J.P. Rice, and D.M. Ginsberg, *Phys Rev Lett* **69** (1992) 824.
2. H. Safar, P.L. Gammel, D.A. Huse, D.J. Bishop, W.C. Lee, J. Giapintzakis, and D.M. Ginsberg, *Phys Rev Lett* **70** (1993) 3800.
3. M. Charalambous, J. Chaussy, P. Lejay, and V. Vinokur, *Phys Rev Lett* **71** (1993) 436.
4. W.K. Kwok, J. Fendrich, S. Fleshier, U. Kelp, J. Downey, and G.W. Crabtree, *Phys Rev Lett* **72** (1994) 1092.
5. H. Pastoriza, M.F. Goffman, A. Arribere, and F. de la Cruz, *Phys Rev Lett* **72** (1994) 2951.
6. E. Zeldov, D. Majer, M. Konczykowski, V.B. Geshkenbein, V.M. Vinokur, and H. Shtrikman, *Nature* **375** (1995) 373.
7. M.A. Manekar, S. Chaudhary, M.K. Chattopadhyay, K.J. Singh, S.B. Roy, and P. Chaddah, *Phys Rev B* **64** (2001) 104416.
8. P. Chaddah, A. Banerjee, and S.B. Roy, Correlating supercooling limit and glass-like arrest of kinetics for disorder-broadened 1st order transitions: relevance to phase separation. http://arxiv.org/pdf/cond-mat/0601095 (2006). Accessed January 5, 2006.
9. K. Kumar, A.K. Pramanik, A. Banerjee, P. Chaddah, S.B. Roy, S. Park, C.L. Zhang, and S.-W. Cheong, *Phys Rev B* **73** (2006) 184435.
10. P. Chaddah, K. Kumar, and A. Banerjee, *Phys Rev B* **77** (2008) 100402.
11. A. Schilling, R.A. Fisher, N.E. Phillips, U. Welp, D. Dasgupta, W.K. Kwok, and G.W. Crabtree, *Nature* **382** (1996) 791.
12. U. Welp, J.A. Fendrich, W.K. Kwok, G.W. Crabtree, and B.W. Veal, *Phys Rev Lett* **76** (1996) 4809.
13. R.M. White and T.H. Geballe, *Long Range Order in Solids*, Academic Press, New York (1979), p. 12.
14. A. Soibel, E. Zeldov, M. Rappaport, Y. Myasoedov, T. Tamegai, S. Ooi, M. Konczykowski, and V.B. Geshkenbein, *Nature* **406** (2000) 283.
15. S.B. Roy, P. Chaddah, V.K. Pecharsky, and K.A. Gschneidner, Jr., *Acta Mater* **56** (2008) 5895.
16. S.B. Roy, G.K. Perkins, M.K. Chattopadhyay, A.K. Nigam, K.J.S. Sokhey, P. Chaddah, A.D. Caplin, and L.F. Cohen, *Phys Rev Lett* **92** (2004) 147203.

17. K.V. Bhagwat and P. Chaddah, *Phys Rev B* **44** (1991) 6950.
18. H. Kuwahara, Y. Tomioka, A. Asamitsu, Y. Moritomo, and Y. Tokura, *Science* **270** (1995) 961.
19. P. Kushwaha, R. Rawat, and P. Chaddah, *J Phys: Condens Matter* **20** (2008) 022204.
20. S. Chaudhary, A.K. Rajarajan, K.J. Singh, S.B. Roy, and P. Chaddah, *Solid State Commun* **114** (2000) 5.
21. S.B. Roy, P. Chaddah, and S. Chaudhary, *Phys Rev B* **62** (2000) 9191.
22. X.S. Ling, S.R. Park, B.A. McClain, S.M. Choi, D.C. Dender, and J.W. Lynn, *Phys Rev Lett* **86** (2001) 712.
23. K.J. Singh, S. Chaudhary, M.K. Chattopadhyay, M.A. Manekar, S.B. Roy, and P. Chaddah, *Phys Rev B* **65** (2002) 094419.
24. M.A. Manekar, S. Chaudhary, M.K. Chattopadhyay, K.J. Singh, S.B. Roy, and P. Chaddah, *J Phys: Condens Matter* **14** (2002) 4477.
25. K.J.S. Sokhey, M.K. Chattopadhyay, A.K. Nigam, S.B. Roy, and P. Chaddah, *Solid State Commun* **129** (2004) 19.
26. M.K. Chattopadhyay, S.B. Roy, and P. Chaddah, *Phys Rev B* **72** (2005) 180401.
27. R. Rawat, K. Mukherjee, K. Kumar, A. Banerjee, and P. Chaddah, *J Phys: Condens Matter* **19** (2007) 256211.

4

Unstable States across First Order Transitions

4.1 Conceptual Difference between Metastable and Unstable States

When liquid is cooled continuously and develops resistance to shear and hardens into the solid state without any signature of a first order transition, glass is formed. The free energy of the glass is higher than that of the equilibrium solid or crystal, and it is at a nonequilibrium state. The decay to the equilibrium solid corresponding to the minimum in free energy could, however, be extremely slow—it may even take hundreds of years—and the glass remains solid over observational timescales.

Crystallization or freezing requires structural rearrangement, and this occurs in a liquid through diffusion. The viscosity of the liquid rises and diffusivity drops with decreasing temperature. This decrease in diffusivity also occurs in a supercooled liquid when it is cooled below its equilibrium freezing (or melting) point. As diffusivity of a supercooled liquid drops, the time for molecular diffusions and rearrangements rises, but it is still able to sample neighboring configurations in the time available. If the time required for the structural rearrangement becomes very large, then the supercooled liquid is said to have vitrified or formed a glass. To put this into perspective, a structural relaxation time of about 100 seconds corresponds to a viscosity of about 10^{13} poise [1] or a diffusivity of 10^{-12} cm^2/second. These limits of viscosity and diffusivity are often used to distinguish a glass from a viscous supercooled liquid. These limits are similar to the limits of resistivity (10^{-12} Ω cm) used in specifying the critical current density J_C of practical superconductors. A different limit for resistivity would give a different value for J_C, but the value changes only slightly as the criterion changes drastically when resistivity varies with current density through a power law with a high index. Similarly, a different limit for viscosity or diffusivity would give a slightly different value for T_g, the temperature at which the liquid forms glass.

In both cases, we are talking of what is referred to as a crossover, not a phase transition. The crossover in both cases is sharp, but it has been studied in great detail in the case of J_C because of the technological implications in superconducting magnets. In that case, the voltage developed (and the flux-flow resistance) depends on J as a power law with index higher than 20 for all technologically relevant superconductors, resulting in the value of J_C depending only very weakly on the resistivity criterion used. If the resistivity criterion is changed even by a factor of 10, the value of J_C changes by only a few percent. This sharp dependence, thus, allows us to use the value of J_C as a material characteristic.

The case for the transformation to glass is similar. If a liquid is cooled below its freezing point, its viscosity also rises at a rapid rate [1]. The viscosity of some liquids rises in accordance with the Arrhenius law:

$$\eta(T) = A \exp(E/kT) \tag{4.1}$$

where A and E are temperature-independent and η rises toward infinity as T drops to zero. η will cross 10^{13} poise at some temperature T_g. This is called the glass transition temperature. Some other liquids exhibit an even more pronounced slowdown with a falling temperature, given by the following Vogel–Tammann–Fulcher (VTF) equation:

$$\eta(T) = A \exp\left[B/(T - T_0)\right] \tag{4.2}$$

Here again, A and B are temperature-independent and η becomes infinite at $T = T_0$. Similar to the widely studied case of flux-flow resistance versus current density in practical superconductors, here also the temperature range over which the transformation in diffusivity or viscosity occurs close to these critical values is narrow and the temperature so identified can be considered an important material characteristic.

We discussed in Chapter 3 how hysteresis can set in if there is a lag in the response because the kinetics is hindered. For the liquid-to-solid phase transition to not proceed, the kinetics must not just be hindered; rather, the transition kinetics must be frozen or arrested. S. Brawer has called glasses liquids with "time held still" [2]. We assert that this can happen even when the state is in a local minimum of free energy. It can happen at $T > T^*$, so that the kinetics is arrested in a supercooled liquid. We shall now be using the term "glass" in this context, i.e., where a thermodynamic first order transition cannot proceed because the kinetics is frozen or arrested. Since the kinetics is frozen, the neighboring configurations cannot be explored. Arrested kinetics overrules thermodynamics, and the glass need not correspond to a local minimum in G(q). T_g can be below or above T^*.

We now assert that a glass can form under quasi-equilibrium slow cooling if T_g is higher than T^*. This is actually not a common situation in structural glasses, where T_g is usually lower than T*. (As noted in Reference 1, T_g is usually around $2T_m/3$, whereas we noted in Section 2.4 that T* is usually around $0.8T_m$.) But a supercooled liquid has slower kinetics than a liquid at the equilibrium freezing point, and it decreases with decreasing temperature. The time required for structural rearrangements rises sharply at lower temperatures, diverging at $T < T_g$. For structural transition to occur, the time-integrated value of the molecular velocity must exceed the length over which structural rearrangements have to take place. Let us consider the water–ice transition. The density of ice is lower by over 10%, and the mean interparticle separation of its H_2O molecules is higher by over 2%. For water to transform into ice, the interparticle separation must rise everywhere by about 0.1 Å: over the length of the sample, the molecules have to move by macroscopic lengths. If we are lowering temperature continuously, then the molecular velocity drops and the total motion possible depends on the time-integral of this continuously decreasing velocity. When the lowest temperature reached corresponds to a near-zero diffusivity or near-zero molecular velocity, then this time-integrated value dictates how far the molecules can move. If this is small, then the molecules cannot reach the equilibrium positions corresponding to the solid ice. It is clear that this time-integral will fall if the cooling rate is high. *The structural transition at freezing can thus be arrested by cooling at a very rapid rate so that this time-integrated value of the molecular velocity is so low that no structural rearrangement could take place.* There is, thus, a critical cooling rate for the formation of glass [3]. The critical cooling rate for glass formation is, hence, another characteristic of a material.

This critical cooling rate is extremely high for metallic glasses, being about 10^6 K/second or higher [3]. If the same cooling were carried out at a rate lower than this critical rate, then crystallization would take place. In these materials where T* is above T_g, there cannot be a smooth transition from the supercooled liquid to glass: any quasi-continuous cooling would convert the supercooled liquid to solid at T = T*. A supercooled liquid can make a smooth transition to glass only if T_g is greater than T*, as is probably the case in glass-formers like O-terphenyl, but it needs extremely rapid cooling if T_g is lower than T*, as in metallic glasses. We shall argue later that in the first order magnetic transitions it is possible to observe both these types of behavior in the same material. The trick involves using appropriate values of magnetic field in which the material is cooled. The first order transition is avoided in both structural and magnetic cases of glass formation, and to prevent a first order transition, the removal of latent heat is also prevented in both cases.

In supercooling, as also in glass formation, crystallization is avoided. Both are metastable states that will relax to the stable state of a crystal. How does a supercooled liquid differ from a glass?

The metastable supercooled state becomes unstable and crystallizes or converts to the equilibrium low T phase below T*, or even at a T closer to the phase transition point T_C, if the fluctuation energy exceeds the height of the barrier $G_B(q)$ around the local minimum at $q = 0$. At the level of folklore, supercooled water will convert to ice under a small disturbance like a pinprick. The pinprick in this folklore provides the fluctuation energy for the metastable liquid to cross the barrier surrounding its local minimum and go to the deeper global minimum. The supercooled liquid is metastable, and the pinprick takes it to the stable state. The glass at T below T* is, on the other hand, thermodynamically unstable because the barrier in $G(q)$ around $q = 0$ has disappeared and it is not in a local minimum of free energy. It does not spontaneously convert to solid, even though the fluctuation energy required for this has become zero. This is because glass is obtained when the kinetics is arrested, and this precludes the possibility of exploring the neighboring structural configurations (in the conventionally used picture) and changing q to reach the nearby minimum in $G(q)$. Thus, the first order transition from liquid to solid is prevented.

The transition from a supercooled liquid to a solid is inhibited by the barrier in free energy. The barrier is higher closer to T_C, when larger fluctuation energy is required. The transition from glass to solid is inhibited by the lack of kinetics, which can be restored by raising the temperature. A supercooled liquid will relax faster as T is lowered and the free energy barrier falls, while a glass will relax faster as T is raised and the kinetics is restored. The maxima and minima in $G(q)$ have become irrelevant in this kinetically arrested state, and that is why glasses are usually visualized as metastable states in potential energy landscape or real space [1]. We are, here, trying to understand the concept of glass, using the ideas of $G(q)$, so that we have a unified way of looking at supercooled states and kinetically arrested states. The glass state may not be thermodynamically metastable because it is not necessary to have a free energy barrier surrounding it. It could even be thermodynamically unstable because the lack of dynamics has overruled thermodynamics. This conceptual distinction is brought out in schematic Figure 4.1.

Second, if we heat a supercooled liquid to above the equilibrium transition temperature, then it reacquires all the characteristics of the equilibrium liquid, including its viscosity. A metallic glass formed by splat-cooling, on the other hand, crystallizes on heating. It would then convert to the equilibrium liquid on further heating but only through a first order transition with discontinuous changes in some properties showing a reentrant behavior. This is depicted in the schematic in Figure 4.2.

We now generalize our discussion beyond the liquid–solid transition to all first order transitions, specifically to magnetic transitions where the potential energy landscape may have no obvious significance. Supercooled states, as also superheated states, are metastable because they are in a local minimum of $G(q)$ and thermodynamically stable against small fluctuations. Kinetically arrested states—and we generalize beyond the one example of a structural

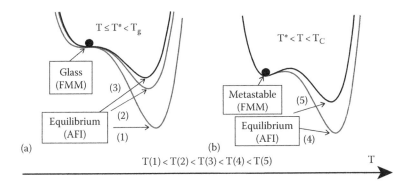

FIGURE 4.1
This schematic shows free energy curves for an arrested glassy state. (Taken from Chaddah, P. et al., *Phys. Rev. B*, 77, 100402, 2008.) The three curves shown in (a) correspond to the higher entropy phase being unstable. The material continues in this phase and does not transform to the global minimum even though there is no barrier preventing this transformation. This is because the kinetics necessary for the transformation is already arrested as $T_g > T^*$. This state is necessarily a glass. In (b), we show the free energy curves for $T > T^*$, where both an arrested state and a metastable supercooled state would persist in the local minimum. As shown, the barrier surrounding the local minimum increases as T rises, and the relaxation rate of the supercooled state (where the kinetics of the transformation is not arrested) would become slower as T rises. But if the kinetics is arrested at $T(4)$ and de-arrested at $T(5)$, then the relaxation rate would become faster as T rises in this range before slowing down above $T(5)$. Thus, a nonmonotonicity in relaxation rate of the metastable phase is a manifestation of T_g being above T^*. In the schematics shown, the glass corresponds to the relaxation rate becoming faster as T rises while the supercooled state corresponds to the relaxation rate becoming slower as T rises.

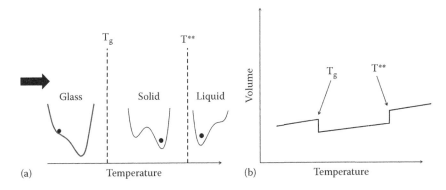

FIGURE 4.2
A glass formed by rapid cooling undergoes two transitions on heating, and its physical properties (such as density) show a reentrant behavior. This is depicted in the schematic in (a). The glass devitrifies to a solid at T_g when the kinetics is de-arrested at $T_g < T^*$. The solid melts into a liquid at T^{**}. Since the glass is formed by rapid quenching of the liquid, this "liquid with time held still" retains some of the liquid's physical properties, such as density. When the glass devitrifies, there is a sharp change in density, just as when the liquid freezes. An approximately reverse jump occurs at T^{**}. This reentrant change in density is depicted in the schematic in (b). We have assumed in the schematic that the solid is denser than the liquid. If the liquid is denser, the reentrant behavior will show a rise at lower temperature followed by a drop.

glass—may or may not be in a local minimum of G(q). Their decay is not inhibited due to the barrier in G(q), and these states can even be thermodynamically unstable. Such kinetically arrested states are like glasses in that they decay at astronomical time scales.

The temperature at which the kinetics slows down to the level that the time required for rearrangements is much longer than the experimental time scale will now be denoted by T_k (as a more general nomenclature of T_g). Since variations in the second control variable (denoted by H) change T_C, we will also consider the temperature for kinetic arrest T_k to be a function of H. We note that T_k does not signify a phase transition and its definition would depend on the "experimental time scale" chosen. However, just as the value of T_g depends only weakly on the critical value of viscosity used as a criterion, so would T_k depend only weakly on the time scale criterion chosen. As noted earlier for T_g, T_k is also a characteristic of the material.

At T*, q = 0 is no longer a minimum in G(q) and the state is not metastable but has become unstable. If T_k < T*, then the higher entropy phase-2 can persist to T_k only on splat-cooling, while slow cooling will cause a conversion to phase-1 at T* > T_k. (Later in this chapter, we shall see how the second control variable can provide a "magic ingredient" to circumvent the need for splat-cooling.) We recognize further that if T_k is larger than T* then we can go by slow cooling from the equilibrium phase-2 to the metastable supercooled phase-2 to a kinetically arrested phase-2 which can persist below T*. This is depicted in the schematic in Figure 4.3. We consider below the temperature dependence of relaxation in the equilibrium state in this case when T_k is larger than T*.

4.2 Relaxation Rates for Metastable, and for Arrested Unstable, States

The kinetically arrested higher entropy phase (or glass)—and also the supercooled phase-2 (or supercooled liquid)—must relax to the crystalline equilibrium state (now being referred to as the lower entropy ordered phase). We have just argued that the relaxation rate of glass will slow drastically as the temperature is lowered—it becomes immeasurable as T drops below T_g. We argue in the following text that the supercooled liquid, or more generally the supercooled higher entropy phase, will relax faster as the control variable crosses the critical value corresponding to the phase transition. We are stressing that order is synonymous with lower entropy and is not necessarily structural order. With these generalizations, we proceed to consider first order phase transitions in general, without restricting ourselves to the liquid–solid transitions.

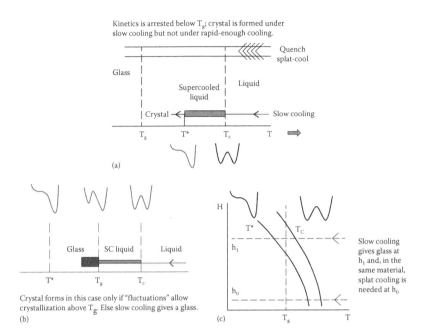

FIGURE 4.3
In (a), we show the situation where a glass is formed by rapid cooling through the freezing point, such that the kinetics of this first order transition is arrested. This happens when $T^* > T_g$ and corresponds to almost all structural first order transitions. The inequality $T^* < T_g$ is depicted in (b) and is commonly encountered in first order magnetic transitions. In this situation, even a slow, quasistatic cooling will result in glass formation. Fluctuations would need to be introduced to cross the barrier in $G(q)$ for the equilibrium crystal to form. Though uncommon in structural glasses, the material O-terphenyl does show formation of a structural glass under slow cooling. In (c), we consider a magnetic first order transition where the higher entropy phase is FM or, more generally, has higher magnetization. The transition temperatures drop with increasing H, with the interesting possibility of the material behaving like O-terphenyl at large H and like a metallic glass at small H.

As we pursue the concepts of a first order transition developed in Chapter 2, we look at free energy as a function of the order parameter with at least two control variables, either of which may drive the first order transition. One control variable is, of course, temperature. In view of the many experimental studies on first order magnetic transitions in a large number of materials, we shall now consider H as the second control variable, with the two phases having distinctly different magnetizations. We note that H can be replaced by P (or E), with FM (or high magnetization phase)/AFM (low magnetization phase) being replaced by high density phase/low density phase (or replaced by high dielectric constant phase/low dielectric constant phase) in this discussion. We also note here that while all control variables influence the height of the barrier separating the metastable and stable states, temperature plays the dominant role in influencing kinetics.

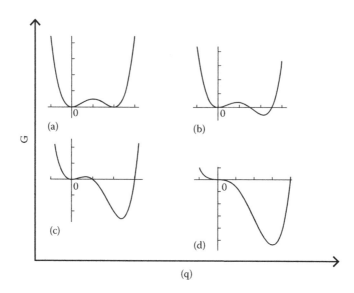

FIGURE 4.4
In (a), we depict G(q) at the phase transition point, while in (b), (c), and (d), we show the schematic after crossing the phase transition point and on reaching the limit of supercooling.

In the schematic Figure 4.4, we show free energy curves as the control variable crosses the phase transition point. Figure 4.4a depicts G(q) as a function of q at the phase transition point, while Figure 4.4b, c, and d show G(q) after crossing the phase transition point and on reaching the limit of supercooling. In (a), q = 0 is an equilibrium state. In (b) and (c), however, the q = 0 state has a barrier hindering the transformation to the ordered phase and can exist, as a supercooled state, even when kinetics exploring the neighboring configurations is allowed. Kinetics is most strongly influenced by temperature, and we consider T as the control variable as we discuss the variation of the relaxation rate.

At T = T_C, the high temperature phase with q = 0 is stable and does not relax. As T is lowered, this supercooled higher entropy phase remains metastable because of the barrier G(q_B). As T falls, the kinetics becomes slower and G(q_B) also reduces. Close to T_C, the enhancement in relaxation due to the drop in the barrier dominates, and relaxation becomes faster as T drops below T_C. As T approaches T_k, the drop in kinetics dominates and the relaxation rate slows down. Relaxation data as T approaches T_k can be fitted well with the Kohlrausch–Williams–Watt stretched exponential function ≈ exp[–$(t/\tau)^\beta$], where τ is the characteristic relaxation time and β is a fitting parameter that typically lies between 0.6 and 0.9 [4]. The characteristic relaxation time τ diverges as T_k is approached from above. In an early study on a first order magnetic transition revealing kinetic arrest, Chattopadhyay et al. [4] measured magnetization of a Ru-doped CeFe$_2$ sample on cooling in various magnetic fields. As mentioned in Chapter 3, this material showed a first order transition from a high temperature ferromagnetic (FM) phase to a low temperature antiferromagnetic (AFM) phase. For a cooling field H = 2 Tesla, there was no measurable relaxation at 55 K, indicating that this temperature is T_C or higher. Relaxation was clearly discernible at 40 K, showing decay from the supercooled FM state to the equilibrium AFM state. The data, reproduced from Reference 4 in Figure 4.5, showed that the relaxation rate increased sharply as T was lowered from 55 to 25 K, with the magnetization dropping by 20% in less than an hour at T = 25 K. The free energy barrier surrounding the supercooled FM state was reducing. The relaxation rate dropped sharply as T was lowered below 22.5, indicating that though the

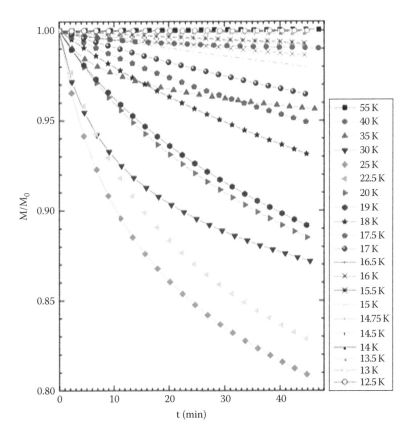

FIGURE 4.5

We show the time decay of magnetization measured at various values of T. (Taken from Chattopadhyay, M.K. et al., *Phys. Rev. B*, 72, 180401, 2005.) All measurements are in H = 2 Tesla, with the Ru-doped $CeFe_2$ sample being cooled from the higher entropy FM phase at 120 K in this fixed field for each T. The solid lines are a fit to the Kohlrausch–Williams–Watt stretched exponential function [4]. The top two curves, showing the slowest relaxation, are at T = 55 K and at T = 12.5 K; the lowest curve showing the fastest relaxation is at T = 25 K.

barrier was reducing, kinetics was reducing faster and the barrier was not being tested. The relaxation rate dropped to half this rate as T was lowered to 19 K, and it became immeasurable at 12.5 K. The supercooled FM state had now been kinetically arrested, as in a glass. This nonmonotonic dependence of the relaxation rate on temperature will be seen when $T_k > T^*$, and this was true in the earlier sample for H = 2 Tesla. This inequality between T_k and T^*, as we shall discuss, does not hold at all values of H.

4.3 Manifestations of Kinetic Arrest in Studies Using Two Control Variables

A very different question that was addressed through these studies on the use of two control variables was the effect of the proximity of the (T^*, H^*) and (T_k, H_k) lines on glass-like arrest of kinetics. It has been well documented in the literature on structural glasses that the critical cooling rate for glass formation falls sharply as T_g/T_m rises [3,5]. It is also conventional wisdom that glass formation is easier as T_g/T_m rises, i.e., as T_m falls and becomes closer to T_g. Empirical data shows that as T_g/T_m rises toward the higher limit of 0.7, the required critical cooling rate is 1 K/second or could even be slower [3,5]. Structural glasses are formed at ambient pressure by using a cooling rate higher than the critical value, and the value of T_g/T_m can only be controlled if T_m drops for a slight change in composition. There has also been the realization that in some materials T_m drops if pressure is raised above ambient [6]. This happens only in those few materials where, like the ubiquitous water, the liquid has a higher density than the solid. Among elements at zero pressure, this is true for Bi, Ga, Ce, Si, and Ge. If one ignores any variation of T_g with pressure, then raising pressure would raise the ratio T_g/T_m in these materials so that the critical cooling rate becomes experimentally achievable [6]. One experiment has been reported where signatures were observed of germanium forming a glass on rapid cooling under pressures of about 8 GPa and higher but not under pressures of 5 GPa or lower [6]. At these higher pressures, T_m dropped enough to make the experimentally realizable cooling rate higher than the critical cooling rate for that T_g/T_m.

We have, however, been discussing the kinetic arrest of a first order transition (or glass formation) by comparing T_k with T^* (and not with T_m or T_C). This is an unconventional view, consequent to our looking at kinetic arrest and glass formation in terms of free energy and not of the potential energy landscape. In Table 2.1, we listed empirical data on the experimentally observed limiting temperature till which supercooling has been observed in various materials. We listed this experimental limit as T^*, and we also listed the melting point T_m. We noted that in all cases T^* is about $0.8T_m$. The conventional wisdom (based on empirical data) on the required critical cooling rate falling to around 1 K/second on T_g/T_m rising toward 0.7 may, thus, correspond to T_k rising toward T*. We have argued in the schematic in Figure 4.3 that if T_k is greater than T* then the kinetics can be arrested even at the slowest possible cooling rate but if T_k is smaller than T* then rapid cooling is essential. *This conclusion is based on physical arguments and not on empirical data. The new understanding attempted was that if T_k was higher than T* then nonergodicity sets in while the system is in the metastable state, and a kinetically arrested glass will form at normal slow cooling rates. If T* was higher than T_k, then the metastable to stable transformation occurs at normal cooling rates and rapid cooling becomes essential for glass-like arrest of first order transition.* We emphasize that the available experimental results for structural arrest of liquids do not contradict this understanding, even though formation of structural glasses has not been specifically analyzed in terms of the supercooling limit T*. As mentioned earlier, this book is about the arrest of first order transitions in general and not restricted to solid–liquid transitions. We are motivated by the immense data on first order magnetic transitions that are now becoming available, but we are conscious that any new concepts we introduce must also hold for the well-studied liquid–solid transitions. Hence, we are confirming that there should be no counter-examples to our concepts.

As T_C depends on the second control variable, so do T* and T**. It follows that the inequality between T* and T_k is a function of the second control variable (which we are considering to be H since the recent data on two control variables is dominated by the second control variable being H) with the further provision that T_k is also a function of H. (We must mention that the pressure-dependence of T_g is not discussed in detail in current literature, beyond the recognition that at very high pressure the high density could cause jamming, resulting in pressure-induced amorphization.)

In Figure 4.6, we consider schematic phase diagrams for first order magnetic transitions. We note that the phase transition line has a slope with the sign being dictated by whether the lower entropy phase has a higher or lower magnetization than that of the higher entropy phase. (In this and all subsequent schematics, the phase transition lines are assumed as straight lines. We recognize that in practical cases there must be a change in slope with changing H, but our phenomenological conclusions depend on the sign of the slope and not on its magnitude. We are discussing new ideas with the simplest approximation, and introducing a curvature does not modify the phenomenology being discussed.) The sign is dictated by Le Chatelier's principle in that an increase in the magnetic field must shift the equilibrium, on the phase transition line on which phase-1 and phase-2 coexist, toward the higher magnetization phase. Increasing H must favor the FM fraction, which in case (a) is phase-2, so an increase in H at fixed T would finally bring us to the right of the transition line where only phase-1 exists. Thus, the transition line must bend to the left if H is increased, and the transition lines must have a negative slope. Similarly in case (b), an increasing H must take us to the

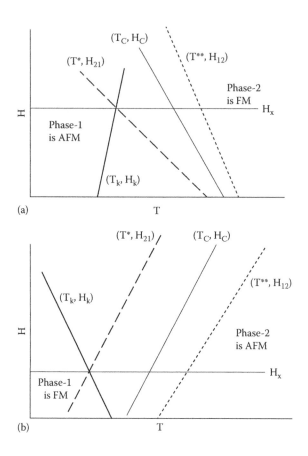

FIGURE 4.6

We show schematics of (T^*, H_{21}), (T_C, H_C), and (T^{**}, H_{12}) lines and also of the (T_k, H_k) line. In (a), the lower entropy phase is AFM, and in (b), the lower entropy phase is FM. The (T^*, H_{21}), (T_C, H_C), and (T^{**}, H_{12}) lines have slopes of the same sign, consistent with the higher magnetization phase being formed as we increase H, or as we increase T. The (T_k, H_k) line has a slope of the opposite sign for the same reason, because the lower temperature phase across this line is the higher entropy and higher magnetization phase-2. The transformation across the (T_C, H_C) line is between two equilibrium phases, those across the (T^*, H_{21}) and (T^{**}, H_{12}) lines are between an equilibrium phase and a metastable phase, and that across the (T_k, H_k) line is between an equilibrium phase and an unstable phase. In (a), the transformation is complete on cooling in $H < H_x$ and arrested on cooling in $H > H_x$. In (b), the transformation is complete on cooling in $H > H_x$ and arrested on cooling in $H < H_x$.

left of the transition line since that is where only FM (phase-1) exists. Thus, the transition line must bend to the right if H is increased, and the transition lines must have a positive slope. The signs of the slopes of (T^*, H_{21}), (T_C, H_C), and (T^{**}, H_{12}) are thus the same and are determined as depicted in Figure 4.6.

We now come to the slope of the (T_k, H_k) line. At temperatures below T_k (which are to the left of this line), the arrested phase-2 exists. Immediately to its right, we are at temperatures above T_k and there is no kinetic arrest, so there we have the equilibrium phase-1. (Much further to the right is the equilibrium phase-2). The arrested phase-2 has retained the properties of the equilibrium phase-2, so in (a) the arrested phase-2 immediately to the left of the (T_k, H_k) line has a higher magnetization than that of the equilibrium phase-1 immediately to the right of the (T_k, H_k) line. It follows that a decreasing H must take us to where the equilibrium phase-1 exists, i.e., to the right of the (T_k, H_k) line. Thus, the (T_k, H_k) line must bend to the right if H is increased, and the (T_k, H_k) line must have a positive slope.

Similarly in (b), the arrested phase-2 immediately to the left of the (T_k, H_k) line has a lower magnetization than that of the equilibrium phase-1 immediately to the right of the (T_k, H_k) line. It follows that an increasing H must take us to the right of the (T_k, H_k) line, where the equilibrium FM phase exists. The (T_k, H_k) line must accordingly bend to the left if H is increased, and the (T_k, H_k) line must have a negative slope. The sign of the slope of the (T_k, H_k) line is thus opposite to that of the (T^*, H_{21}), (T_C, H_C), and (T^{**}, H_{12}) lines.

These arguments give us the sign of the slopes but not their magnitudes. The magnitude for the (T_C, H_C) line can be obtained from the latent heat using the Clausius–Clapeyron relation. We do not have a corresponding prescription to obtain the slopes of the (T^*, H_{21}), (T^{**}, H_{12}), and (T_k, H_k) lines. Nevertheless, *we have come to a crucial conclusion that the slope of the (T_k, H_k) line is opposite to that of the supercooling limit (T^*, H_{21}). This would help us, as a "magic ingredient" referred to in the previous section, to vary the relative values of Tk and T*. We can thus create situations in which $T_k > T^*$ and a first order transition can be arrested even under slow cooling.*

4.3.1 Temperature Variations in Various Fields

Let us now consider the schematic Figure 4.6a corresponding to a high-T higher entropy phase with high magnetization (or density or dielectric constant) that makes a first order transition to a lower entropy phase with low magnetization. The most common example of this is the FM to AFM transition, and many such magnetic transitions have been studied in the past two decades. In most cases, the FM to AFM ordering causes a remapping of the Fermi surface because of the appearance of zone boundaries (following an increase in the unit cell), which results in a sharp increase in the resistivity. The most extensively studied transition of this type is the charge ordering transition in half-doped manganites that has been discussed earlier in this book and that is accompanied by a metal to insulator transition with a very large rise in electrical resistance. This first order transition is manifested by a sharp drop in the thermodynamic property of the magnetization and by a sharp rise in the nonequilibrium property of the resistivity. In Figure 4.7, we show such measurements of both magnetization and resistance in various magnetic materials. A hysteresis between the cooling and warming cycles is clearly seen in all cases. Curves (a) and (b) are for $La_{0.5}Ca_{0.5}MnO_3$ (LCMO) [11], curves (c) and (d) are for Co-doped Mn_2Sb [14], and curves (e) and (f) are for Al-doped CeFe$_2$ [7–9]. Most of the resistance data shown is in H = 0; magnetization cannot be measured in H = 0 and the data shown is in low fields, where H is below the field H$_x$ where T$_k$ and T* cross.

The first order transition from a high magnetization phase-2 to a low magnetization phase-1 is, thus, easy to detect. Let us now look at how kinetic arrest influences this transition. When cooling from a high temperature at low values of H, we cross T* before T$_k$ is reached. At cooling rates that are below some "critical cooling rate" the phase space is explored, and we assume that we are below this critical cooling rate so that thermodynamics prevails and a transition to the low entropy AFM phase is made. On further

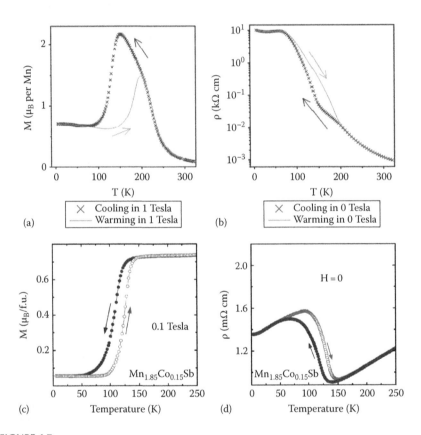

(a)

(b)

(c)   Temperature (K)

(d)   Temperature (K)

FIGURE 4.7

This shows data for the phase transition depicted in Figure 4.6a, where the lower entropy phase is AFM. This has a higher resistance and a lower magnetization than those of the higher entropy FM phase. As depicted in Figure 4.6a, this transition is completed on cooling in a low H that is smaller than H$_x$. We show data from our measurements, of both magnetization and resistance, in various materials. Curves (a) and (b) are for La$_{0.5}$Ca$_{0.5}$MnO$_3$ (LCMO) taken from Reference 11, and curves (c) and (d) are for Co-doped Mn$_2$Sb taken from Kushwaha et al. [14]. All magnetic fields used are below the relevant H$_x$, and the forward and reverse curves show a hysteresis.   *(Continued)*

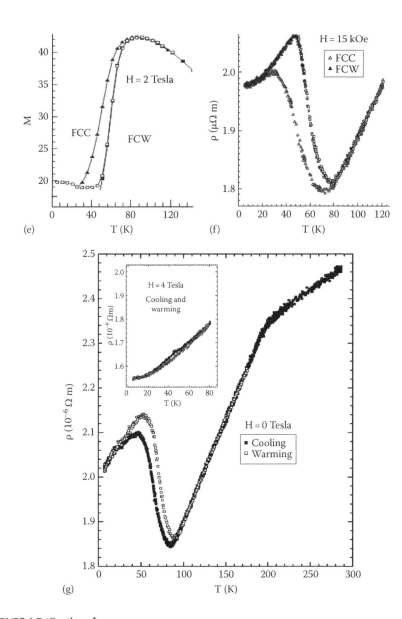

FIGURE 4.7 (*Continued*)
In curves (e) and (f), we show the original data on Al-doped $CeFe_2$ taken from References 7–9 at H below H_x. Curve (g), taken from Singh et al. [9], shows that the transition is clearly seen in resistance measurements in H = 0 and is totally absent in resistance measurements in H = 4 Tesla. These works on Al-doped $CeFe_2$ [7–9] initiated the idea that coexisting phases observed in half-doped manganites were due to kinetic arrest of the first order transition.

cooling, T_k will be encountered, but this has no observable effect because the material is already in a global minimum of free energy. The situation for $T_k < T^*$ is similar to that of a metallic glass, where the first order transition can only be arrested under rapid cooling. In the rest of the chapter, we will assume that the cooling is slow or quasistatic so that the first order transition can be arrested only if T^* is below T_k.

Let us now consider what happens if the material is cooled in a large H. As argued in Section 4.3, $T_k(H)$ rises and $T^*(H)$ decreases with an increasing H. We assume that this cooling field is high enough such that $T^*(H)$ is lower than $T_k(H)$. The material is now in a local minimum of $G(q)$ at $T_k(H)$, and so is in the supercooled high entropy FM phase when the kinetics is arrested. As it is cooled toward $T^*(H)$, the barrier in free energy reduces and finally vanishes but kinetic arrest prevents it from exploring phase space: the FM phase is now unstable, but its kinetics is arrested. This prevents it from transforming to the AFM phase. This is clearly seen in Al-doped $CeFe_2$ [9], as shown in Figure 4.7g, where the resistive transition is kinetically arrested on cooling in H = 4 Tesla.

The state at the low temperature (T_1, H_0) is obtained by cooling in a low magnetic field such that $T_k(H_0)$ is lower than $T^*(H_0)$, while that at (T_1, H_1) is obtained by cooling in a high magnetic field such that $T_k(H_1)$ is higher than $T^*(H_1)$ (see the schematic in Figure 4.8a). The two states are different in that the former is an equilibrium state while the latter is a kinetically arrested state. We list some experimental observations of this behavior in the following text. The first such observation attributed to kinetic arrest was in Al-doped $CeFe_2$, where the transition was not observed when H_1 was above 4 Tesla. The arrest of the transition at such fields was reported in magnetization measurements by Manekar et al. [7,8] and in resistance measurements [7,9]. This was then followed by an observation of the kinetic arrest in Ru-doped $CeFe_2$, where magnetization measurements [4] and resistance measurements [10] showed that the transition was arrested when H_1 was above 3 Tesla. This has been followed by a large number of similar reports in various materials. To list a few, arrest was observed in LCMO on cooling in fields above 7 Tesla [11,12], in Co-doped Mn_2Sb on cooling in 8 Tesla [13], in Pd-doped FeRh on cooling in 8 Tesla [14], and in Co-doped NiMnSn alloy on cooling in 9 Tesla [15]. In each of these cases, the higher entropy phase had higher magnetization and was FM (except in the case of Co-doped Mn_2Sb where it was ferrimagnetic but still had higher magnetization) and the lower entropy phase was AFM. In all cases, measurements were made during both cooling and warming in various fixed fields, and for the lower values of cooling field when the transition was observed, a thermal hysteresis was also clearly observed in the transition. The magnetic transition was always first order. In all cases, the behavior of resistance followed that of magnetization. A sharp rise in resistance together with a sharp fall in magnetization was observed on cooling in low fields, and hysteresis was observed between the cooling and heating cycles. The resistive and magnetic transitions were not seen on cooling in large fields, consistent with the kinetic arrest of a first order transition.

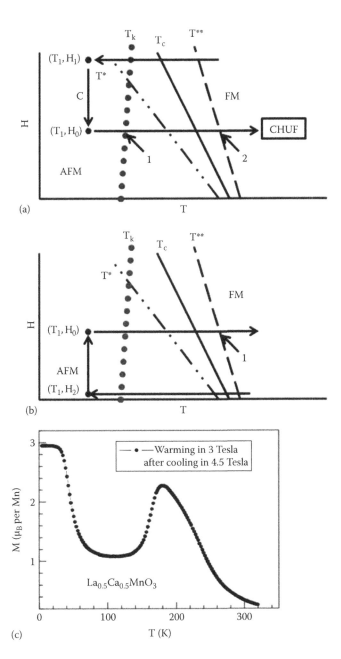

FIGURE 4.8
In (a) and (b), we show schematics for the CHUF protocol for an AFM lower entropy phase. In (c), we show the reentrant behavior observed by us in LCMO [11] for the path corresponding to (a).

4.3.2 CHUF Protocol

In Figure 4.8a, we cool the higher entropy FM phase to T_1 in the constant field H_1, with $T_k(H_1) > T^*(H_1)$, such that the transformation to AFM phase is arrested. We then lower H from H_1 to H_0 with T_1 being below T_k at all H. The kinetics then remains frozen because T does not rise above $T_k(H)$ at any H. The field H_0 is such that $T_k(H_0) < T^*(H_0)$, and the equilibrium AFM state is obtained by lowering the temperature in the fixed field H_0. In Figure 4.8b, we cool to T_1 in a field H_2, which is lying below H_0, such that the AFM state is obtained. We then raise the field to H_0 and thus have a path-dependent state at (T_1, H_0). Let us now start raising the temperature with the field fixed at H_0. (We have named this protocol CHUF, for "cooling and heating in unequal fields." The full potential of this new protocol will be brought out in Chapter 5.)

If we reach H_0 from H_2, then we do not see any change on crossing T_k since the material is already in the equilibrium state. We do, however, observe a sharp rise in magnetization at the temperature $T^{**}(H_0)$ as it transforms to the higher entropy FM phase. If, on the other hand, we reach H_0 from H_1, then we observe a sharp drop in magnetization at $T_k(H_0)$ because the kinetically arrested high magnetization (and low resistance) phase gets de-arrested and transforms to the equilibrium AFM phase. As we raise T further, we then observe a sharp rise at $T^{**}(H_0)$. We, thus, observe a reentrant transition, as the temperature is raised in this case. Measured data corresponding to the reentrant transition is depicted in Figure 4.8c. The CHUF protocol thus provides a drastic visual signature through two sharp transitions: a glass-like arrested state transforming to the low entropy equilibrium phase (as in devitrification of a glass) at $T_k(H_0)$ and then transforming to the high entropy equilibrium phase (as in melting of a solid) at $T^{**}(H_0)$. The resistance measurement would show a sharp rise at $T_k(H_0)$ followed by a sharp decrease at $T^{**}(H_0)$. This protocol also provides a means for measuring $T_k(H_0)$. Since we can vary H_0, this CHUF protocol allows determination of this kinetic arrest line in the entire region where T_k is lower than T^*. Such measurements, as have been reported in various materials, shall be discussed later. We just assert here that the measurements of $T_k(H)$ have shown slopes that are consistent with what we argued as following from Le Chatelier's principle, namely, the values rise with an increasing H when the lower entropy phase is AFM and decrease when the lower entropy phase is FM.

Let us now consider the schematic Figure 4.6b corresponding to a high-T higher entropy phase with low magnetization (or density or dielectric constant) that makes a first order transition to a lower entropy phase with high magnetization. In all these cases, T_C rises as the second control variable rises. The most common example of this is an AFM to FM transition with decreasing temperature. Such magnetic transitions have also been studied in the past two decades. Let us examine how the kinetic arrest influences this transition.

When cooling from a high temperature, as before, but at high values of $H = H_2$, we cross T^* before T_k is reached. Since we assume cooling rates that are below the "critical cooling rate," the configuration space is explored and a transition to the low entropy FM phase is made. On further cooling, T_k is encountered, but this has no observable effect because the material is already in a global minimum of free energy. Let us now consider what happens if the material is cooled in a low value of $H = H_1$ such that $T^*(H_1)$ is lower than $T_k(H_1)$. The material is in the supercooled high entropy AFM phase when the kinetics is arrested. As it is cooled to $T^*(H_1)$, the barrier in free energy vanishes and the higher entropy phase is now unstable, but the kinetic arrest prevents it from transforming to the lower entropy phase. The states at $T_1(H_1)$ and $T_1(H_2)$ are different, in that the former is a kinetically arrested state while the latter is an equilibrium state. In contrast to the case of Figure 4.6a, we now have a situation where a first order AFM to FM transition, with a thermal hysteresis, should be observed in the higher values of H but will be kinetically arrested at values below critical value of H (say H_k). This value H_k corresponds to $T_k(H_k) = T^*(H_k)$. Measurements showing this kind of behavior were first reported in Gd_5Ge_4 by Magen et al. [16], where a hysteretic transition was observed on cooling and heating in H = 5 Tesla, while no transition was observed in H = 0. Roy et al. [17] then reported an AFM to FM transition on cooling, with thermal hysteresis, in a single crystal of Gd_5Ge_4 persisting to a low cooling field of H = 1 Tesla. A very extensively studied material in this class was Al-doped PCMO, for which Banerjee et al. [18] showed both magnetic and resistive transitions that were observed with a clear thermal hysteresis in high cooling fields but were arrested at low fields below 1 Tesla. Other materials in which this behavior has been reported are Ta-doped $HfFe_2$ [19] and Tb-doped LCMO [20]. The hysteresis observed in magnetization and resistance measurements on varying T in high H are depicted in Figure 4.9 for Al-doped PCMO [18].

Let us explore how the CHUF protocol works in this case. We choose a field $H_0 < H_k$ lying between H_1 and H_2 and a temperature T_1 that is below T_k for all $H < H_2$. We can access $T_1(H_0)$ by either raising the field isothermally from H_1 or lowering it isothermally from H_2. The latter path will give us the equilibrium FM state, while the former path will give us the arrested AFM state. Thus, we get two different states at the same point $T_1(H_0)$, depending on the paths followed. Let us now start raising the temperature with the field fixed at H_0. If we reach H_0 from H_1, then we observe a sharp rise in magnetization at the temperature when we cross T_k and the arrested AFM phase is de-arrested and transforms to the equilibrium FM phase. On further heating, the material will transform to the higher entropy AFM phase at T^{**}, with a reentrant drop in magnetization. If, on the other hand, we reach H_0 from H_2, then we observe no change as we cross T_k because the material is already in the equilibrium lower entropy phase. On further heating, we observe a sharp drop in magnetization at T^{**} as the material transforms to the higher entropy phase. These are depicted in the schematic Figure 4.10. The CHUF protocol now provides a reentrant transition when the cooling field is low, in contrast to the earlier case when a reentrant transition was seen when the cooling field was high.

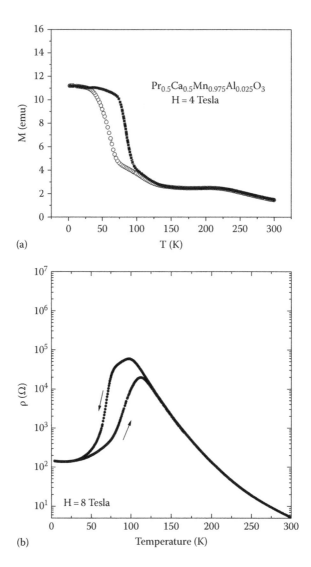

(a)

(b)

FIGURE 4.9
In (a) and (b), we show the magnetization and resistance measurements on a varying T in a constant H for a FM lower entropy phase, based on data we published earlier in Reference 18. The hysteresis is clearly observed.

4.3.3 CHUF for Supercooled States

Let us now consider the case where the state at $T_1(H_1)$ in the case of Figure 4.8a or that of Figure 4.10a was a supercooled state and not an arrested state. This happens if kinetic arrest has not occurred, as is depicted in Figure 4.11. We note that the behaviors depicted in Figures 4.8 and 4.10, showing a reentrant transition on heating in constant H, cannot happen in such cases. The supercooled

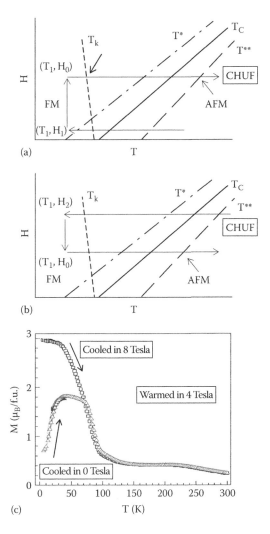

FIGURE 4.10
In (a) and (b), we show the counterparts of Figure 4.8 for a FM lower entropy phase. In (c), we show M(T) on warming in 4 Tesla, based on data that we published earlier [18]. A reentrant transition is seen when the cooling field (0 Tesla) is lower than the warming field.

state will transform to the lower entropy equilibrium state during the isothermal change of H, and the warming process completes the reverse transition to the higher entropy state. The absence of a reentrant transition on warming provides a check that the metastable state is a supercooled state and not a kinetically arrested state. This is experimentally a much easier test than the test for temperature dependence of relaxation rate.

The observations of a reentrant transition under the CHUF protocol provide a striking visual confirmation that one of the two paths resulted in a

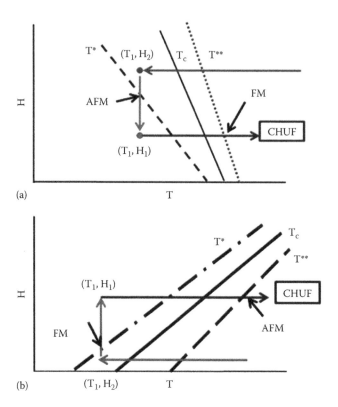

FIGURE 4.11
We consider that kinetic arrest does not occur, and we have a supercooled metastable state at (T_1, H_2). We follow the protocols described in Figures 4.8 and 4.10, and the forward transition is completed during the isothermal variation of H, but no reentrant transition is seen during warming.

kinetically arrested state, and this brings out the potential of this protocol. A reentrant transition on heating, for some appropriate values of the cooling and heating fields, is therefore a clear signature of a first order transition having been arrested. All physical properties that change sharply at the phase transition would show this reentrant behavior. As we shall discuss in Chapter 5, this reentrant behavior has been observed in many materials exhibiting a first order magnetic phase transition.

4.3.4 Isothermal Variations of the Field

We have discussed the behavior under varying T in constant H, as we look at the interplay between supercooling and kinetic arrest using two control variables. We shall now discuss the behavior on varying H in constant T. We stress that measurements under isothermal variations in a magnetic field, commonly referred to as hysteresis measurements, are the most common measurements that magnetic materials are subjected to. The schematics in Figure 4.12 depict

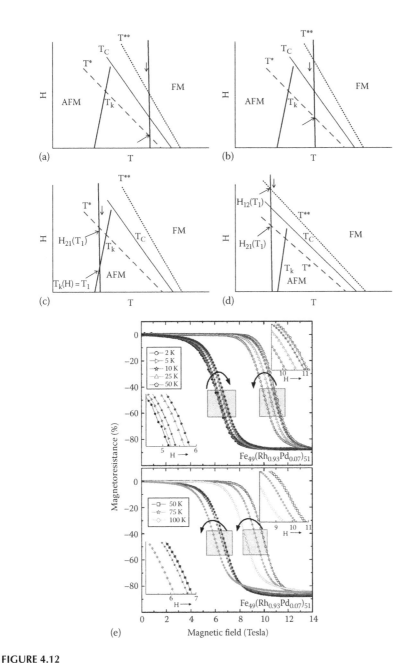

FIGURE 4.12

We consider an isothermal decrease in H from a large value of H. An FM to AFM transition is observed as H is reduced. The temperature (for this isothermal reduction of H) is being lowered in (a)–(d). The value of H at which the FM to AFM transition is seen increases from (a) to (b) but decreases beyond (c). This is observed in our data on Pd-doped FeRh in figure (e). The FM state remains arrested in (d), and we have the virgin curve outside the hysteresis envelope in an M-H or an R-H measurement. (e: Taken from Kushwaha, P. et al., *Phys. Rev. B*, 80, 174413, 2009.)

an isothermal measurement protocol that is used to measure magnetization hysteresis curves or resistance hysteresis curves. As mentioned earlier, magnetization or resistance can be replaced by any physical property that changes sharply between the two phases. In schematic Figure 4.12, the lower entropy phase is AFM while the higher entropy phase is FM.

As we cool from a high temperature in $H = 0$, the higher entropy FM phase transforms to the lower entropy AFM phase at T^*. When lowering the temperature further to below T_k, the kinetics is arrested but this has no measurable effect as there are no further transformations. We now start raising H until we cross $H_{12}(T_1)$, at which field the phase transforms into the equilibrium higher entropy FM phase. We now start lowering H. In the absence of a kinetic arrest, it should transform to the AFM phase at $H_{21}(T_1)$, i.e., when it crosses the T^* line.

This is the case for the schematics in Figure 4.12a and b, but consider the case for the schematic in Figure 4.12d. Here, T is below T_k and the transformation to the equilibrium FM phase on reducing H is arrested. The material remains in the arrested FM phase as H is cycled isothermally. So the virgin curve, when we started raising H after cooling in zero field, corresponds to the AFM phase whereas the remaining cycles all correspond to the FM phase. This gives rise to the interesting phenomenon of the virgin curve lying outside the envelope curve. This observation was first made by Manekar et al. [7] in isothermal measurements of both resistance and magnetization in Al-doped CeFe$_2$, and it has since been reported in more than a dozen families of magnetic materials. As Manekar et al. explained in a subsequent paper [8], they had carefully chosen a material where disorder-broadening of the first order transition—a topic we shall discuss in detail in Chapter 5—is large. This virgin curve lying outside the envelope curve being observed as T is lowered is a signature of kinetic arrest of a first order transition, subject to some cross-checks by doing similar measurements at different temperatures. We shall now discuss the behavior at higher values of T.

In Figure 4.12c, we consider a temperature T_1 lying slightly above $T_k(H = 0)$. The initial state is of a lower entropy AFM phase, which transforms to a higher entropy FM phase at $H_{12}(T_1)$. In the field reducing cycle, it does not transform back to AFM at $H_{21}(T_1)$ because the temperature is below T_k at that field. It transforms back only at the lower value of field H_L, satisfying $T_k(H) = T_1$. We no longer have the virgin curve lying outside the envelope, but we have an interesting feature that the back transformation occurs at a field H_L that increases with the rising temperature. At any higher temperature that is greater than the temperature at which T_k and T^* lines cross, the kinetic arrest will not influence the hysteresis envelope and the back transformation will occur at $H = H_{21}$. As in Figure 4.12a and b, H_{21} then falls with rising temperature. This nonmonotonic behavior of the reverse transformation was studied in some detail by Kushwaha et al. [24] in Pd-doped FeRh, as shown in Figure 4.12e.

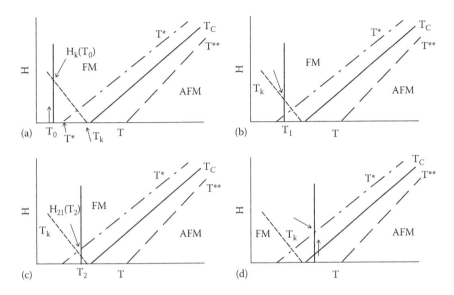

FIGURE 4.13

We consider an isothermal increase in H after lowering T in H = 0. An AFM to FM transition is observed as H is raised. The temperature (for the isothermal increase in H) is being increased in (a)–(d). The value of H at which the AFM to FM transition is seen reduces from (a) to (b) but increases from (c) to (d). This is observed in our data on Al-doped PCMO that is published in Reference 23.

We shall now follow through this discussion for the case of schematic Figure 4.13, which corresponds to the higher entropy phase being AFM and the lower entropy phase being FM. We choose for the isothermal measurement a temperature T_0 that is lower than $T^*(H = 0)$, which is the lowest value of T^* in this case and is lower than T_k at this H = 0. We shall consider the behavior for higher values of T later.

As we cool from a high temperature in H = 0, the higher entropy AFM phase should transform to the lower entropy FM phase at T^*, but the kinetics is arrested at the higher temperature of T_k and the material remains in the arrested higher entropy phase at T_0. We now start raising H until we cross the T_k line at $H_k(T_0)$, at which field the phase transforms into the equilibrium lower entropy FM phase. As T_k falls with rising H, the field at which the transition is observed falls with rising T_0. We now start lowering H. The material remains in the equilibrium FM phase as H is cycled isothermally. So the virgin curve, when we started raising H after cooling in zero field, corresponds to the AFM phase, whereas the remaining cycles all correspond to the FM phase. This again gives rise to the interesting phenomenon of the virgin curve lying outside the envelope curve. However, here the virgin curve (for magnetization, resistance, or any physical property that changes sharply between the FM and the AFM phases) corresponds to the kinetically arrested higher entropy phase and the envelope curve corresponds to the equilibrium lower entropy phase. The qualitative nature here is the same as that for the earlier discussion of a lower entropy AFM phase, and the visually drastic observation of the virgin curve lying outside the envelope curve is also seen here. This observation for a lower entropy FM phase was first made by Chattopadhyay et al. [21] in magnetization measurements on Gd_5Ge_4 at $T_0 = 5$ K, and it has since been reported in many magnetic materials showing an AFM to FM transition on cooling. The very striking similarity for the two cases of AFM and FM low T phases was highlighted by Banerjee et al. [22] when they juxtaposed the measured M-H curves for LCMO and for PCMAO at 5 K. *The virgin curve lying outside the envelope hysteresis curve does not determine what the ground state is.* CHUF measurements clarified that the LCMO has an AFM state with lower entropy, while PCMAO has an FM state with lower entropy. We must assert here that, as discussed in Section 2.6 and shown in Figure 2.10, the virgin curve can sometimes lie outside the envelope curve even without kinetic arrest. This stresses the importance of the CHUF protocol for detecting kinetic arrest.

We shall now discuss the behavior at higher values of T, where differences appear between the two cases. At a temperature T_1 lying above $T^*(H = 0)$ but below $T_k(H = 0)$ depicted in (b), the initial state is of higher entropy AFM phase. This is in an arrested state. As H is raised, it crosses the T_k line but is below the T^* line and converts to the lower entropy FM phase. The field at which the AFM to FM transition is observed decreases with increasing T_1 since it is dictated by the $T_k(H)$ line. At a higher temperature T_2, depicted in (c), it crosses the T_k line while above the T^* line. The AFM state is now de-arrested but is a metastable supercooled state since T_2 is above T^* at this point. It will convert to the stable FM phase on crossing the T^* line. The field at which the AFM to FM transition is observed now rises with increasing temperature, as also seen in (d), since it is now dictated by the $T^*(H)$ line, whereas at lower T it was dictated by the $T_k(H)$ line. The slopes of these two lines are opposite and the T-dependence of the forward (increasing H) AFM-FM transition shows a nonmonotonic dependence on T. This is observed in our data on Al-doped PCMO that is published in Reference 23.

To summarize the results of this section, the use of two control variables (T and H for the case of first order magnetic transitions) allows experiments to distinguish a kinetically arrested state, which is not sitting in a local minimum of G(q), from a supercooled metastable state, which is sitting in a local minimum of G(q). Using the CHUF protocol, the kinetically arrested state would show a reentrant transition, for a range of monotonically changing values of the second control variable (H) fixed during cooling (H_{Cool}), evolving into a single transition at the higher temperature end after crossing a certain value of the second control variable. (The supercooled state will not show this reentrant behavior.) When the transition temperature rises as the value of the second control variable is raised, this evolving into a single transition will be observed as the value of H_{Cool} is raised. When the transition temperature drops as the value of the second control variable is raised, this evolving into a single transition will be observed as the value of H_{Cool} is lowered. This result has been confirmed experimentally for first order magnetic transitions with H as the second control variable, but it is also applicable for other second control variables such as pressure and electric field.

The protocol discussed above showed visually different behaviors as T was raised. The second distinctive behavior is observed by isothermal variation of the second control variable at different values of T. At very low T, where a state is kinetically arrested, the virgin curve for any physical property that varies sharply between the two phases lies outside the envelope curve. The values of the second control variable, at which the forward transition and the reverse hysteretic transition are seen, vary with temperature T. If the state is kinetically arrested, then one of these two values shows a nonmonotonic variation with T. When the transition temperature rises as the value of the second control variable is raised, this nonmonotonic variation is in the forward transition, but if the transition temperature drops as the value of the second control variable is raised, then this nonmonotonic variation is in the reverse transition. Again, this result has been confirmed experimentally for first order magnetic transitions with H as the second control variable, but it is also applicable for other second control variables such as pressure and electric field.

In Chapter 5, we shall discuss many experimental observations. This discussion will be taken up after we introduce disorder-broadened transitions. Most studies of first order magnetic transitions are on materials with possible functionalities. These possibilities are enhanced in multicomponent materials with enhancing substitutions, where disorder is inherent to such materials.

References

1. P.G. Debenedetti and F.H. Stillinger, *Nature* **410** (2001) 259.
2. S. Brawer, *Relaxation in Viscous Liquids and Glasses*, The American Ceramic Society, Columbus, OH (1985).
3. A.L. Greer, *Science* **267** (1995) 1947.
4. M.K. Chattopadhyay, S.B. Roy, and P. Chaddah, *Phys Rev B* **72** (2005) 180401.
5. A. Inoue, *Acta Mater* **48** (2000) 279.
6. M.H. Bhat, V. Molinero, E. Soignard, V.C. Solomon, S. Sastry, J.L. Yarger, and C.A. Angell, *Nature* **448** (2007) 787.
7. M.A. Manekar, S. Chaudhary, M.K. Chattopadhyay, K.J. Singh, S.B. Roy, and P. Chaddah, *Phys Rev B* **64** (2001) 104416.
8. M.A. Manekar, S. Chaudhary, M.K. Chattopadhyay, K.J. Singh, S.B. Roy, and P. Chaddah, *J Phys: Condens Matter* **14** (2002) 4477.
9. K.J. Singh, S. Chaudhary, M.K. Chattopadhyay, M.A. Manekar, S.B. Roy, and P. Chaddah, *Phys Rev B* **65** (2002) 094419.

10. K.J.S. Sokhey, M.K. Chattopadhyay, A.K. Nigam, S.B. Roy, and P. Chaddah, *Solid State Commun* **129** (2004) 19.
11. P. Chaddah, K. Kumar, and A. Banerjee, *Phys Rev B* **77** (2008) 100402; A. Banerjee, K. Kumar, and P. Chaddah, *J Phys: Condens Matter* **20** (2008) 255245.
12. S. Dash, K. Kumar, A. Banerjee, and P. Chaddah, *Phys Rev B* **82** (2010) 172412.
13. P. Kushwaha, R. Rawat, and P. Chaddah, *J Phys: Condens Matter* **20** (2008) 022204.
14. P. Kushwaha, A. Lakhani, R. Rawat, A. Banerjee, and P. Chaddah, *Phys Rev B* **79** (2009) 132402.
15. A. Banerjee, S. Dash, A. Lakhani, P. Chaddah, X. Chen, and R.V. Ramanujan, *Solid State Commun* **151** (2011) 971.
16. C. Magen, L. Morellon, P.A. Algarabel, C. Marquina, and M.R. Ibarra, *J Phys: Condens Matter* **15** (2003) 2389.
17. S.B. Roy, M.K. Chattopadhyay, P. Chaddah, J.D. Moore, G.K. Perkins, L.F. Cohen, K.A. Gschneidner, Jr., and V.K. Pecharsky, *Phys Rev B* **74** (2006) 012403.
18. A. Banerjee, K. Mukherjee, K. Kumar, and P. Chaddah, *Phys Rev B* **74** (2006) 224445.
19. R. Rawat, P. Chaddah, P. Bag, P.D. Babu, and V. Siruguri, *J Phys: Condens Matter* **25** (2013) 066011.
20. R.R. Doshi, P.S. Solanki, U. Khachar, D.G. Kuberkar, P.S.R. Krishna, A. Banerjee, and P. Chaddah, *Physica B* **406** (2011) 4031.
21. M.K. Chattopadhyay, M.A. Manekar, A.O. Pecharsky, V.K. Pecharsky, K.A. Gschneidner, Jr., J. Moore, G.K. Perkins et al., *Phys Rev B* **70** (2004) 214421.
22. A. Banerjee, K. Kumar, and P. Chaddah, *J Phys: Condens Matter* **21** (2009) 026002.
23. K. Mukherjee, K. Kumar, A. Banerjee, and P. Chaddah, *Eur Phys J* **B86** (2013) 21.
24. P. Kushwaha, A. Lakhani, R. Rawat, and P. Chaddah, *Phys Rev B* **80** (2009) 174413.

5

Disorder-Broadened Transitions

5.1 Broadened Supercooling and Superheating Bands

As mentioned in Chapter 4, most magnetic materials of interest are multi-component systems with doping and substitutions, with an intrinsic frozen disorder. Early theoretical arguments of Imry and Wortis [1] showed that disorder would result in a rounding of a first order transition, with the broadened transition remaining first order for a small disorder. Since the correlation length for a first order transition is finite, the transition proceeds through nucleation of the finite regions of the second phase. It can be understood that these regions, with dimensions of the order of correlation length, can have slightly varying "compositions" because of the disorder. Timonin [2] considered such a case of a lattice divided into blocks having different transition temperatures and found a gradual transformation from a homogeneous disordered state to a homogeneous ordered one. These finite regions have slightly varying transition temperatures (or transition value for the other control variable), and this results in a spatial distribution of the T_C, T^*, and T^{**} lines across the sample. This spread of the local T_C values across the samples would result in the T_C, T^*, and T^{**} lines being broadened into a band, with each line in the band corresponding to a different local region. The transition between the two phases occurs over a range of T (or H), and the two phases grow (or diminish) as this range is traversed. The state of each local region is dictated by how far away from the ambient T and H the T_C and H_C of that region are. The coexisting phases collapse into a single homogeneous phase as this broadened (T_C, H_C) band is exited. For such a broadened transition, it becomes difficult to distinguish the latent heat from a peak in specific heat.

In an extremely pure system, the first order transition is sharp, as shown by the thick dashed line in Figure 5.1a. Such a system can be supercooled through careful experimentation, and in that case, the transition would follow the solid line marked by T_3-T_4-T^*. Similarly, the superheated phase would follow the solid line T_1-T_2-T^{**}. Such transitions occur with the control parameter not changing its value, i.e., the entire sample transforms at the same temperature (homogeneous nucleation). If the transition is diffuse, the superheated and

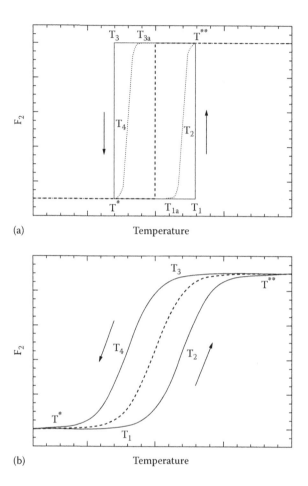

(a)   Temperature

(b)   Temperature

FIGURE 5.1
We depict schematics for the fraction of higher entropy phase-2 as a function of temperature for various levels of disorder-induced broadening. (Taken from Chattopadhyay, M.K. et al., *Phys. Rev. B*, 68, 174404, 2003 [3].) In (a), the solid lines are the T* and T** for a sharp transition in a pure material and the dotted lines are for slight broadening due to disorder. In (b), the disorder has caused a high level of broadening. Hysteresis is seen in all cases. In the cases shown in (a), T_1, the temperature at which the transition starts on heating, is higher than T_3, the temperature at which the transition starts on cooling, but in case (b), T_1 is lower than T_3.

supercooled phases would follow the dotted curves T_{1a}-T_2-T^{**} and T_{3a}-T_4-T^*, respectively. The hysteretic first order transition would occur over a range of values, as depicted in the schematic Figure 5.1a. Such diffuse transitions can result from impurities and disorders in the sample and represent material characteristics. (They can also have artifact-based origins, such as the demagnetization effects related to the sample shape. In the very interesting case of a spherical sample of a Type-I superconductor in a uniform magnetic field, the effective field in the polar region is much smaller than that in the equatorial region. The resulting phenomenon, of an intermediate state that smoothens the superconducting-normal transition, is beyond the scope of this book.)

In the case of the slight broadening depicted in Figure 5.1a, the transition during cooling is initiated at a temperature lower than the temperature at which the transition is initiated during warming. The hysteresis due to supercooling and superheating in materials without disorder, as discussed in the previous chapters, also follows this behavior. We note that the transition can even be broadened so much, as in Figure 5.1b, that the transition during cooling is initiated at a temperature higher than the temperature at which the transition is initiated during warming. We shall discuss later some actual observations of hysteretic transitions that are broadened so much.

The first visual realization of such a landscape of local T_C values, establishing a broad first order transition with an increasing magnetic field in single crystals of the superconductor $Bi_2Sr_2CaCu_2O_8$, was provided by Soibel et al. [4] for the vortex melting transition. For isothermal scans, the differential magneto-optical technique was used to image vortex structures with a spatial resolution of about 6 μm. At the temperature of 70 K, vortex solid to vortex liquid transition was found to be initiated on the field rising to 88.5 Oe, and it was completed on the field increasing to 99.5 Oe. A similar visual realization of a landscape of local T_C values resulting in a broad antiferromagnetic (AFM) to ferromagnetic (FM) transition in Ru-doped $CeFe_2$ was provided by Roy et al. [5], using scanning Hall probe imaging with a sensor of dimensions 5 μm × 5 μm. They studied the first order transition being caused by variations of temperature in a field H = 3.5 Tesla, where they observed a broad transition along with hysteresis. In the heating cycle, the transition from AFM to FM state was initiated at about 40 K and was completed at about 70 K, while in the cooling cycle, the reverse transition was initiated at about 60 K and completed below 30 K. It was clear that the transition during cooling is initiated at a temperature that is higher than the temperature at which the transition is initiated during warming. The hysteresis, observed even at the mesoscopic scale with the micro-Hall probe, emphasizes that the transition is still of first order. Roy et al. [5] also reported measurements in isothermal scans of H and showed data for the temperature held fixed at 60 K. They showed a broad transition with a similar visual realization of a landscape of local H_C values under the variation of the field. In the increasing-H cycle, the transition from AFM to FM state was initiated at about 2 Tesla and completed at about 4 Tesla, while in the reducing-H cycle, the reverse transition was initiated at about 3.5 Tesla and completed below 1.5 Tesla. Again, onset on increasing H was at a lower field than onset on decreasing H. The hysteresis was again confirmed visually even at the mesoscopic scale with the micro-Hall probe images, as depicted in Figure 5.2.

In all three cases described, hysteresis was clearly observed at this mesoscopic length scale at which the second phase was nucleating. The major qualitative deviation this caused in phenomenology, from both the Ehrenfest classification and the modern classification, is that *disorder-broadened*

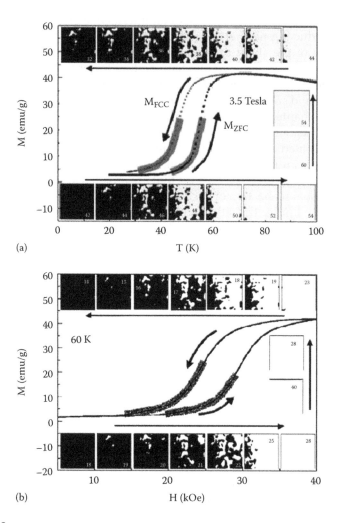

(a)

(b)

FIGURE 5.2

We show in (a) our data on Ru-doped $CeFe_2$ for the first order transition being caused by variation of temperature in a field H = 3.5 Tesla. In the heating cycle, the transition from AFM to FM
state, as observed through bulk magnetization measurements, was initiated at about 40 K and
completed at about 70 K, while in the cooling cycle, the reverse transition was initiated at about
60 K and completed below 30 K. Hysteresis was also observed using scanning Hall probe imaging with a sensor of dimensions 5 μm × 5 μm; photographs of the scanned images are shown
at various temperatures. The dark regions correspond to the AFM phase. In the heating cycle,
the transformation starts at 44 K while in the cooling cycle it is completed only below 32 K. The
observation of hysteresis at the mesoscopic scale, with the micro-Hall probe, emphasizes that
the transition is first order. We also reported measurements in isothermal scans of H, and data
for temperature held fixed at 60 K shown in (b) (Taken from Roy, S.B. et al., *Phys. Rev. Lett.*, 92,
147203, 2004). The hysteresis observed through bulk magnetization measurements is shown,
and the scanning Hall probe images confirm that the FM phase starts forming at 1.9 Tesla with
an increasing H but converts fully to AFM phase below 1.5 Tesla with a decreasing H.

first order transition cannot support the concept of an observable latent heat. The transition no longer being sharp implies that the entropy changes over a finite range of the control variable. In the space of two control variables, we no longer have lines on which the two phases coexist and on which the phase transition occurs—we now have bands. In Figure 5.3a, we depict the schematic for the (T_C, H_C) lines corresponding to the phase transition, as also (T^*, H^*) lines and (T^{**}, H^{**}) lines corresponding to the limits of metastability. In Figure 5.3b, we show the broadened bands, but when the forward transition is initiated at a higher value of the control parameter and the reverse transition at a lower value, as in Figure 5.1a, the bands do not overlap. When the transition becomes so broad that the forward transition is initiated at a lower value of the control parameter and the reverse transition at a higher value, as in Figure 5.1b, then the bands overlap, and the schematic in Figure 5.3c depicts a confusing picture. To maintain clarity, we shall pursue our discussion using Figure 5.3b. We emphasize that the much higher broadening indicated in Figures 5.1b and 5.3c is frequently encountered, and this is discussed in the following text.

The concept of T^* and T^{**} bands implies that the transition between two magnetic phases would occur over a range of T (or H). Instead of the two phases coexisting at one temperature, they now coexist across this band of temperatures (or of H). Consistent with the case of vortex lattice melting, and with the case of doped $CeFe_2$, we consider first the case where the higher entropy phase has larger magnetization and the lower entropy phase has smaller magnetization. Each of T^*, T_C, and T^{**} then lower with increasing H, and this is what was depicted in the schematic in Figure 5.3a through c. The fraction of the smaller magnetization phase would grow as this range is traversed by lowering T. The state of each local region is dictated by how far away from the ambient T and H the T_C and H_C of that region are. The coexisting phases would collapse into a single homogeneous phase as this broadened band is exited. The phase fraction of the two phases, as the control variable crosses the transition value, was depicted in schematic Figure 5.1. In the absence of disorder-induced broadening and with no fluctuations, a sharp transition should occur at the limit of supercooling and that of superheating. With disorder broadening, we have these occurring over a range of values of the control variable. The supercooling and superheating lines are now replaced by a T^* band (or H_{21} band) and a T^{**} band (or H_{12} band), and these bands are well separated as indicated in the schematics in Figure 5.3b. The T_C or H_C band lies between these two. We shall assume this small broadening to maintain clarity in our discussion, but we must assert that experiments show that the transition can remain first order even at much higher levels of broadening, as depicted in Figures 5.1b and 5.3c, and as shown in the data in Figure 5.2. As mentioned earlier, the disorder-induced broadening of first order magnetic transitions is seen in magnetic materials with a potential for applications, and even in single crystal samples.

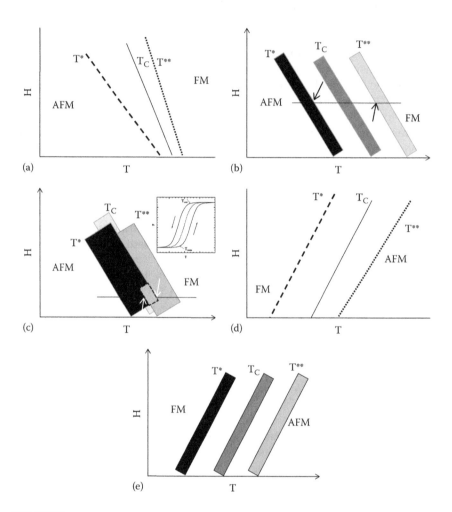

FIGURE 5.3
We depict in (a) the transition temperature T_C and the metastability limits T^* and T^{**} for a first order magnetic transition where the lower entropy phase is AFM and the higher entropy phase is FM. Each of these temperatures decreases with increasing H whenever the lower entropy phase corresponds to a lower magnetization state. In (b), disorder has broadened each of these temperatures from a line to a band in (H, T) space. The transition on heating is initiated on entering the T^{**} band, and the transition on cooling is initiated on entering the T^* band and this occurs at a lower temperature. These temperatures are indicated (for one particular value of H) by arrows, and this inequality is as for the case (a) without disorder broadening. In (c), we consider a high level of disorder broadening as in Figure 5.1b. The initiation on cooling is now at a higher temperature than that on warming, at various values of H. In (d) and (e), we show the counterparts of (a) and (b) for the case when the lower entropy phase is FM or, more generally, the lower entropy phase corresponds to a higher magnetization state.   (*Continued*)

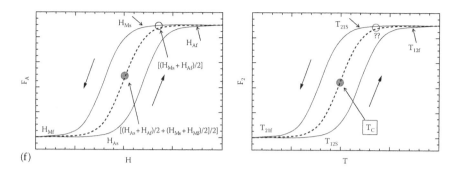

FIGURE 5.3 (*Continued*)
In (f), we show schematics for a very broad martensite to austenite transition with increasing field and for a transition from phase-1 to phase-2 with increasing temperature. The transition point has to be defined carefully, as discussed in the text.

There has always been some ambiguity in defining the transition temperature for an experimentally observed broad transition in a physical property. The frequently used criterion locates where the change in slope is the largest. Another often used criterion attempts to locate the starting point T_s and the finishing point T_f of the transition. These are commonly identified as the points where 10% and 90% of the transition takes place. The transition is then taken to occur at $[T = (T_s + T_f)/2]$. This criterion is commonly used when a transition is observed over a finite range. The situation is more complicated when the first order transition shows hysteresis, as we now measure the starting and finishing values for the control variable on the supercooling and superheating limits. The increasing-H cycle would then be used to estimate H_{12} as $(H_{12s} + H_{12f})/2$. The decreasing-H cycle should similarly be used to estimate H_{21} as $(H_{21s} + H_{21f})/2$.

Problems occurred in studies on FM shape memory alloys (FSMA) where attempts were made to extract entropy differences between the martensite and austenite phases. It is necessary to have a proper estimate of a quantity like entropy, especially when it is being estimated to detect the origin of a new physical phenomenon like kinetic arrest [6–9]. The control variable for this FOMT transition is H, and these works infer the transition point $H_C = [(H_{21s} + H_{12f})/2]$ as the critical value for the phase transition [6–9]. (In their nomenclature, they use $H_C = (H_{Af} + H_{Ms})/2$, where H_{Ms} is the starting point of the transition to the lower entropy martensite phase and H_{Af} is the ending point of the transition to the higher entropy austenite phase.) From the schematic in Figure 5.3f, it is obvious that this will always overestimate H_C. Similarly, for isothermal scans, the transition temperature where the two free energies are equal is estimated as $T_C = [(T_{Af} + T_{Ms})/2]$ or, in our nomenclature, $[(T_{21,s} + T_{12,f})/2]$. It is again obvious that this will overestimate T_C. While $H_{21,f}(T)$ does show anomalous downward curvature at low temperatures in these reports, the $H_C(T)$ obtained by the procedure mentioned earlier showed a flattening at low T. This inferred flattening was then used to estimate a vanishing entropy difference between the two phases and then infer a possible origin of kinetic arrest.

A better estimate to H_C is obtained from the average of the midpoints of the increasing- and decreasing-H cycles, and we get $H_C = [(H_{12,s} + H_{12,f})/2 + (H_{21,s} + H_{21,f})/2]/2$, as was used by Lakhani et al. [10] for the control variable being T. They obtained an $H_C(T)$ that does show an anomalous downward curvature, instead of a flattening one, at low temperatures. We attributed this downward curvature to an interplay of supercooling and kinetic arrest [10]. We emphasize that experiments may not directly measure $H_C(T)$, especially for disorder-broadened first order transitions.

We now examine how this disorder broadening is experimentally influenced by the presence of a kinetic arrest.

5.2 Kinetic Arrest and Interrupted Transitions: Tuning by Cooling Field

We have discussed earlier the scenario where (T^*, H^*) and (T_k, H_k) lines cross, bringing out the evolution from a metglass-like to an O-terphenyl-like situation. With (T^*, H^*) becoming a band, this evolution has a new intermediate regime where T_k at some H falls within the T^* band. This is depicted in the schematic in Figure 5.4a, corresponding to the lower entropy phase having lower M or lower density. For a field H below a certain H_L, T_k is lower than the entire T^* band. Above a certain H_U, T_k is higher than the entire T^* band. In these two cases, there is no qualitatively new feature beyond that discussed in Chapter 4. For H lying between H_L and H_U, we now have T_k crossing the T^* band. Let us cool slowly, or quasistatically, in such an H. The low-T phase starts forming when T is lowered, in a field H lying between H_L and H_U, to a value $T_U(H)$ where it enters the T^* band at that H. The first order transition proceeds with phase-1 forming increasingly till T_k, when it is arrested or interrupted (see schematic Figure 5.4a). We make the simplifying linear approximation, that the fraction of phase-1 increases linearly with decreasing T. This allows us to introduce our new idea without complicated algebra and also without loss of generality.

We, thus, have the two phases coexisting and growing until the temperature equals T_k at that H. No further transformation from phase-2 to phase-1 now occurs as T is below $T_k(H)$. The two phases coexist with these phase fractions, and this coexistence persists all the way as T is lowered below T_k with the fraction of phase-2 being given in the linear approximation by $X_2 = [T_k - T_L]/[T_U - T_L]$ where $T_L(H)$ is the low temperature end of the T^* band. Since each of T_K, T_U, and T_L is a function of H, X_2 is obviously a function of the cooling field H.

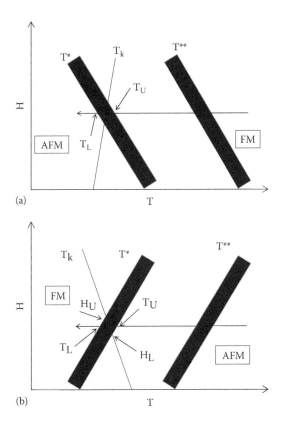

(a)

(b)

FIGURE 5.4
We consider in (a) that the lower entropy phase is AFM, and in (b) that the lower entropy phase is FM. These are counterparts of the schematics in Figure 4.6 but with each of T*, T_C, and T** being broadened into bands. We do not show the T_C band in this and the subsequent figures for reasons of visual clarity; we also assume that the transformation occurs at the limits of supercooling and superheating. Here, the T_K line overlaps with the T* band at some H, and the transformation is interrupted after it has proceeded in part, i.e., up to $[T_U - T_K]/[T_U - T_L]$ (which is obviously a function of the field H). The arrested fraction, or the fraction of coexisting phases, can thus be tuned by the cooling field.

In the schematic in Figure 5.4b, we consider the lower entropy phase having a higher M or higher density. For a field H below a certain H_L, T_k is higher than the entire T* band, while above a certain H_U, T_k is lower than the entire T* band. We can follow through the arguments in the earlier paragraph, and we again get $X_2(H) = [T_k(H) - T_L(H)]/[T_U(H) - T_L(H)]$ where $T_L(H)$ and $T_U(H)$ are the low and the high temperature ends of the T* band at that value of field.

The corresponding dependences of X_2 on cooling field are depicted in Figure 5.5a and b, respectively. These are in contrast with the scenario without disorder broadening. As discussed in Chapter 4, for the lower entropy phase having lower magnetization or density, X_2 is either 1 or 0, depending

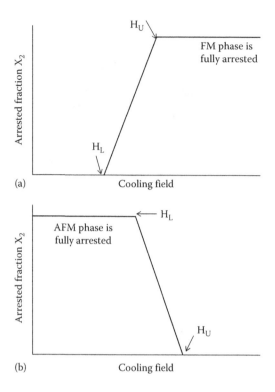

FIGURE 5.5
The corresponding dependences of the arrested fraction X_2, on the cooling field, are depicted in (a) and (b), respectively. Here, the arrested fraction increases (or reduces) with increasing value of the cooling field for the AFM (or FM) lower entropy phase. X_2 can be continuously varied between 0 and 1 by varying the cooling field between H_L and H_U. This is in striking contrast with the scenario without disorder broadening, where the fraction changes abruptly from 0 to 1 as the cooling field is changed.

on whether the cooling field is lower or higher than the value of H_X where T* and T_K lines intersect. In this case, X_2 is zero when the material is cooled in a field below H_L and is 1 when the material is cooled in a field H_U or higher. Phase coexistence being obtained with the cooling fields lying between H_L and H_U is a qualitatively new result, which is a consequence of the first order transition becoming broad. This was a drastic concept explaining the phase coexistence observed in the half-doped CMR manganites as persisting to the lowest T [11], which would not violate any basic law because it referred to states out of equilibrium. This was unlike the concept of an inhomogeneous ground state, which raised the possibility of finite entropy at 0 temperature in an equilibrium state.

We must recognize for consistency that we are discussing regions with varying T_C. Why, then, should they have the same T_k? If such a kinetic arrest were to occur below a (H_k, T_k) line in the pure system, would not the disordered system have a (H_k, T_k) band formed out of the quasi-continuum of (H_k, T_k) lines where each line would correspond to a local region of the sample? This was first addressed by us in a series of papers [11–14]. This broadening will result in other new observations, and we shall discuss this after completing the discussion on broadened T* and T** bands. Postponing the discussion on T_k bands, we continue with a line for (H_k, T_k) and with bands for T* and T** (or H_{21} and H_{12}) for estimating the arrested phase fraction $X_2(H)$.

Direct experimental determination of the arrested phase fraction $X_2(H)$ is difficult because magnetization values are not linearly proportional to the phase fraction. This is because magnetization in both high entropy and low entropy phases is H-dependent. As an extreme example, magnetization is zero at H = 0 in both phases. (This extreme example is not true for resistivity, which remains different for the two phases even at H = 0.) Both FM and AFM phases do show magnetoresistance, but the relation between resistance and phase fraction is also not linear. Nonetheless, both these physical properties are different in the two phases at most (H, T) points, and this provides some measures that can approximately estimate X_2. If their values at a fixed (H, T) point show a dependence on the value of the field in which the sample was cooled to T, then this would confirm that X_2 is dependent on the cooling field. This control of X_2 by choosing different cooling fields is an easily observable consequence of kinetic arrest across a disorder-broadened first order transition.

The phase fraction being tunable with the cooling field was a test of the new concept of an interrupted first order transition—a concept we proposed some years after the possibility of an inhomogeneous ground state was considered to explain the observation of low temperature phase coexistence in half-doped CMR manganites [15,16]. Our concept was introduced while presenting our studies on doped $CeFe_2$, which showed interesting thermomagnetic history effects in our extensive magnetotransport and magnetization measurements across the FM to AFM phase transition. A comparison was made with the comparatively limited studies of thermomagnetic history effects by Kuwahara et al. [15] across the FM to AFM transition in $R_{0.5}Sr_{0.5}MnO_3$ (RSMO) crystals, which we then explained as a similar phenomenon. We predicted that similar features would be seen if more extensive studies were made on RSMO.

We studied this doped $CeFe_2$ material (which was of mainly academic interest with no potential for applications, in contrast to the potential of CMR manganites for applications) extensively and persistently for a few years to understand and test the validity of our new concept and to explain the data of Kuwahara et al. [15]. Magen et al. [17] followed our concept in their study of thermomagnetic history effects in Gd_5Ge_4, a magnetocaloric material with potential for applications, and explained their observations with our model and again contrasted them with the data on RSMO crystals. Once our extensive study on doped $CeFe_2$ was done to a certain degree of satisfaction and peer acceptance, we followed new protocols to report and explain unusual observations in materials that were being competitively studied by many groups because of their potential for application. Measurements following conventional protocols had, we argued, masked the equilibrium phase diagram because the arrest of dynamics masked thermodynamics. We now discuss some of these new protocols that test this proposal of a disorder-broadened first order transition being interrupted midway and before completion. In keeping with the phase diagrams of doped $CeFe_2$ and RSMO we shall first consider in detail the case of the lower entropy phase being the lower magnetization AFM phase. The case of the lower entropy phase being the higher magnetization FM phase shall be discussed separately and analogously.

After cooling in a field H to a temperature below T_k at that H, we can now cool further with intermediate changes in H but always staying in the hatched region at temperatures below the $T_k(H)$ line so that no further transformation is possible. Because the states are kinetically arrested, there will be no change in the fractions of the two coexisting phases that is dictated by the cooling field in which $T_k(H)$ line was crossed. Thus, at any point (H_0, T_0) below the $T_k(H)$ line, i.e., $T_0 < T_k(H_0)$, the two phases can be made to coexist with an arbitrary ratio of the two fractions by choosing an appropriate cooling field. Any desired ratio X can be obtained at such (H_0, T_0) point by cooling to below the $T_k(H)$ line in an H such that $X = [T_k(H) - T_L(H)]/[T_U(H) - T_L(H)]$ and then varying H and T to reach (H_0, T_0). In this process, one must not warm to above $T_k(H)$, i.e., one must not exit the shaded region shown in Figure 5.6. As X varies with a cooling field, physical properties would also vary.

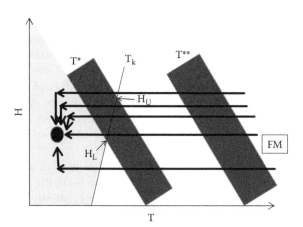

FIGURE 5.6

We show five different paths to obtain various values of the arrested FM fraction at the same point. The lower entropy phase is AFM. Any traversal within the shaded region does not change X_2. The path shown for the lowest cooling field gives a fully AFM state, while that in the highest field gives a fully arrested FM state. The other three paths give a partially arrested FM phase coexisting with an equilibrium AFM phase.

This procedure was followed by Banerjee et al. [18] in various half-doped manganites to obtain continuously variable values of magnetization and resistivity at the same value of H and (low value) of T. For PSMO, we first showed that phase-2 is FM with low resistance and phase-1 is AFM with high resistance. This was established through observation of a hysteretic change in magnetization when cooling and warming in H = 1 Tesla. We then cooled the sample to 5 K in different values of H and then varied H isothermally to the same value H_0 after each cooldown. In Figure 5.7, we show results for H_0 = 4 Tesla. M-H measurements showed that even in the fully FM phase, the magnetization is below the saturation value ($3.5\mu_B$) at this field. The measured M at H = 4 Tesla saturates to about $3\mu_B$ for a cooling field 4 Tesla and drops monotonically to about $1\mu_B$ for a cooling field of 0 Tesla. (This is the value of M at 4 Tesla in the fully AFM phase. Even the resistance drops by a factor of 4 in the fully AFM-I phase as H is raised from 0 to 4 Tesla.) The FM fraction is clearly rising as the cooling field is rising, and this is further confirmed by measurements of resistance done in H = 4 Tesla. As inferred from the conductance data in Figure 5.7, the resistance drops with increasing cooling field and saturates only at a cooling field of about 6 Tesla. From the data shown, one can conclude that at T = 5 K, H_L is about 1 Tesla and H_U is about 6 Tesla. Similar measurements of M were done with H_0 = 1 Tesla by varying the cooling field from 0 to 8 Tesla and resistance measurements were done with H_0 = 0 by

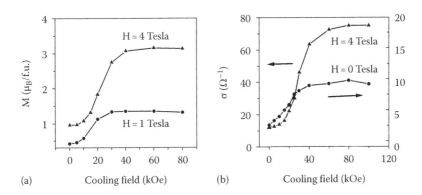

(a) Cooling field (kOe)   (b) Cooling field (kOe)

FIGURE 5.7
We show in (a) the values of magnetization at a fixed temperature of 5 K and fixed magnetic fields of 4 Tesla (upper curve) and 1 Tesla (lower curve). Different values are obtained after cooling in different fields (shown on x-axes) to 5 K and then isothermally varying H to the measurement value. Our figure for $Pr_{0.5}Sr_{0.5}MnO_3$ (PSMO) sample, which has a lower entropy AFM state. In (b), we show values of conductivity at a fixed temperature of 5 K and fixed magnetic fields of 4 Tesla (upper curve) and 0 Tesla (lower curve). Different values are obtained after cooling in different fields (shown on x-axes) to 5 K and then isothermally varying H to the measurement value. The figure is again for PSMO sample, which has an AFM and high resistance (low conductivity) lower entropy state. (Taken from Banerjee, A. et al., *J. Phys. Condens. Matter*, 18, L605, 2006.)

varying the cooling field from 0 to 10 Tesla, and measured values varied monotonically with cooling field. Specifically, the magnetization at 5 K in $H_0 = 1$ Tesla could be obtained as higher (by cooling in H > 4 Tesla) than that at $H_0 = 4$ Tesla (by cooling in H < 1 Tesla). Similarly, the resistance at $H_0 = 0$ at 5 K could be made lower than at H = 4 Tesla under suitably chosen paths for reaching these two points. This is counterintuitive because we expect to observe a larger FM phase in a larger H, favoring higher magnetization and lower resistance. These new results can be understood by the existence of a disorder-broadened T* band crossing a kinetic arrest line T_k, resulting in interrupted first order transitions. We emphasize that a cooling-field dependence of the phase fractions (or of X_2) was first observed in the widely studied "potential functional material" RSMO and not in doped $CeFe_2$, but the exhaustive study along various paths in H-T space was based on physics ideas that were developed and refined in the "academic material" $CeFe_2$.

This use of cooling field to tune the fraction of coexisting phases was again emphatically shown by Lakhani et al. [10] by studying the magnetization in a magnetic shape memory alloy. They deduced the two phase fractions by fitting the magnetization obtained in a measuring field of 4 Tesla to the M-H curves of the two pure phases. They could vary the arrested austenite fraction from 0% to 80% by varying the cooling field from 0 to 8 Tesla.

We shall consider in the following sections further predictions as this state of coexisting phases is heated under different H, and as the applied field is varied isothermally, and contrast these with the predictions under the "cooling and heating in unequal field" (CHUF) protocol made in Chapter 4.

Before that, however, we must consider the case where the higher entropy phase is AFM and the lower entropy phase is FM. When the higher entropy phase is FM, we discussed that the transformation from a metglass-like situation to an O-terphenyl-like situation, in the same material, happens with rising H as $T_k < T^*$ changes to $T_k > T^*$. When the lower entropy phase is a FM, the change from a metglass-like situation to an O-terphenyl-like situation happens with a reducing H. We discuss this by using the schematic in Figure 5.8. The largest value of T_k is at H = 0, and the smallest value of T* is also at H = 0. So if we have the physically significant situation of $T_k > T^*$, it must hold at H = 0 and T_k and T* will cross at some higher value of H. This was already emphasized in Chapter 4. We now have the requirement that at H = 0 some part of the T* band must lie below T_k. For the sake of simplicity in discussion, Figure 5.8 assumes that the entire T* band lies below T_k at H = 0, so that the AFM to FM transition on cooling is completely arrested at H = 0. T_k starts crossing the lower end of the T* band at H_L and lies completely above the T* band at H_U. This was presented in the schematics in Figures 5.4b and 5.5b, and we discuss it in detail in the following text.

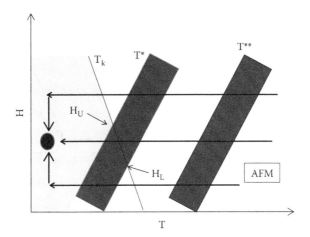

FIGURE 5.8
We show different paths to obtain various values of the arrested AFM fraction at the same point. The lower entropy phase is FM. Any traversal within the shaded region does not change X_2. The path shown for the lowest cooling field gives a fully arrested AFM state, while that in the highest field gives a fully FM state. The third path in a field intermediate between H_L and H_U gives a partially arrested AFM phase coexisting with an equilibrium AFM phase.

If the material is cooled in a field lower than H_L, then T_k is crossed before the T* band is reached. The AFM to FM transition would be totally arrested, and the material will persist in the AFM phase (or phase-2) to the lowest temperature. If the material is cooled in a field between H_L and H_U, then we have T_k crossing the T* band. The transition from phase-2 to phase-1 is initiated at $T_U(H)$ and would have been completed at $T_L(H)$, but it proceeds only until it is interrupted at $T_k(H)$. We, thus, have the two phases coexisting and growing until the T equals T_k at that H. No further transformation from phase-2 to phase-1 now occurs as T is below $T_k(H)$. As discussed earlier, we get $X_2(H) = [T_k(H) - T_L(H)]/[T_U(H) - T_L(H)]$, where $T_L(H)$ and $T_U(H)$ are the low temperature and high temperature ends of the T* band at that value of the field. The two phases now continue to coexist as T is lowered below T_k. X_2 is 1 for $H < H_L$ and reduces monotonically to 0 at $H = H_U$. This was emphasized with schematic figures in studies on the manganite Al-doped $Pr_{0.5}Ca_{0.5}MnO_3$ (PCMAO) by Banerjee et al. [18], as well as in studies on $La_{0.215}Pr_{0.41}Ca_{0.375}MnO_3$ (LPCMO) by Wu et al. [19]. We show the schematic Figure 5.8, in which the $T_k(H)$ line lies above the T*(H) band at H = 0 and starts intersecting with this band at finite fields.

For the PCMO sample, we first showed that phase-2 is AFM with high resistance and phase-1 is FM with low resistance through a hysteretic

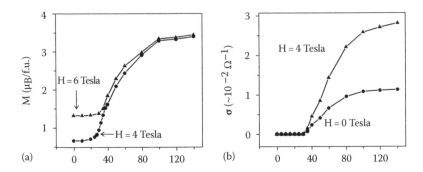

FIGURE 5.9
We show our data for Al-doped $Pr_{0.5}Ca_{0.5}MnO_3$ sample, which has a lower entropy FM state. (Taken from Banerjee, A. et al., *J. Phys.: Condens. Matter*, 18, L605, 2006.) The curves in (a) show various values of magnetization measured at a fixed temperature of 5 K and fixed magnetic fields of 6 Tesla (upper curve) and 4 Tesla (lower curve). In (b), we show values of conductivity measured at a fixed temperature of 5 K and fixed magnetic fields of 4 Tesla (upper curve) and 0 Tesla (lower curve). Different values are obtained at the same (H, T) point after cooling from 320 to 5 K in different fields (shown on x-axes) and then isothermally varying H to the measurement value. This data, and that shown in Figure 5.7, firmly established that coexisting phase fractions could be continuously tuned by using different cooling fields.

change in magnetization when cooling and warming in H = 1 Tesla. We then cooled the sample to 5 K in different values of H to obtain coexisting phases with different phase fractions. Having obtained coexisting phases with a certain X_2, we kept this fraction fixed as we traversed within the hatched region of (H, T) space that lies below the $T_k(H)$ line. This is similar to the case of the lower entropy phase being AFM. So we then varied H isothermally to the same value H_0 after each cooldown. In Figure 5.9, we show results for H_0 = 4 Tesla. The measured M at H = 4 Tesla saturates to about $3.5\mu_B$ for the cooling fields above 10 Tesla, indicating a fully FM phase. It drops monotonically to about $0.5\mu_B$ for cooling fields below about 2.5 Tesla, indicating a fully AFM phase. For magnetization measured in H_0 = 6 Tesla, we get about $1.25\mu_B$ for cooling fields below about 3.5 Tesla, and it saturates to about $3.5\mu_B$ for cooling fields above 10 Tesla. Further, the FM fraction is clearly rising monotonically as the cooling field is rising, and this is confirmed by measurements of resistance done in H_0 = 4 Tesla. As inferred from the conductance data in Figure 5.9, the resistance starts dropping as the cooling field rises above 3 Tesla and saturates only at a cooling field of above 10 Tesla. From the data shown, one can conclude that at T = 5 K, H_L is about 2.5 Tesla and H_U is about 10 Tesla. Similar measurements of resistance were done in H_0 = 0 Tesla by varying the cooling field from 0 to 12 Tesla, and measured values reduced monotonically with increase in cooling field. Specifically, the magnetization at 5 K in $H_0 = 4$ Tesla could be obtained as higher than that at $H_0 = 6$ Tesla under suitably chosen paths for reaching the two points. Similarly, the resistance at H = 0 Tesla at 5 K could be made lower than that at H = 4 Tesla under suitably chosen paths for reaching these two points. This is counterintuitive because we expect that a larger measuring field should favor the FM metallic phase. These new results can again be understood by the existence of a disorder-broadened T* band crossing a kinetic arrest line T_k, resulting in interrupted first order transitions.

To conclude this discussion, if we are cooling to a low temperature that lies below $T_k(H)$ with T_k lying above the T* band at some H, then for both cases of FM and AFM states as the lower entropy state depicted in Figures 5.6 and 5.8, cooling in H = 0 will result in a pure AFM phase while cooling in large H will result in a pure FM phase. Cooling in a band of intermediate fields lying between H_L and H_U will result in phase coexistence down to the lowest temperature, with the fraction of the FM phase rising from 0 to 1 as this band is traversed. The data in Figures 5.7 and 5.9 show that there is a striking qualitative similarity between the two contrasting lower entropy states. More careful studies that look for thermal hysteresis do, however, bring out a clear distinction between the two different lower entropy states. In Figure 5.10, we show measurements on Al-doped PCMO sample, which has a lower entropy FM state, taken from reference [20]. The resistance data in (a) is measured on cooling in different H. It shows that transition is initiated at a higher temperature as H rises, but T_k is above the T* band at H = 0. It follows the general statement made earlier that cooling in higher field gives a lower resistance corresponding to a larger fraction of the FM metallic phase. There is a sharp drop in resistance indicating that the phase transition results in a lower entropy phase that is metallic. The magnetization data in (b) is taken during cooling and warming, and shows an AFM to FM transition that is hysteretic, and where the magnitude of the jump reduces as H is reduced consistent with a larger fraction of the higher entropy AFM phase being arrested.

In Figure 5.11, we show data on three different materials whose lower entropy state has lower magnetization and higher resistance. Similar to the data in Figure 5.10, all cases show thermal hysteresis with the jump having a magnitude that does depend on the cooling field H. Consistent with the discussion on FM to AFM transition, all materials show that the transition is being arrested in large values of H.

Figures 5.10 and 5.11, thus, help distinguish the nature of the lower entropy state. This clearly indicates the new phenomenon of a disorder-broadened first order transition that is being interrupted by kinetic arrest. It is shown that the transition can be interrupted at different stages of

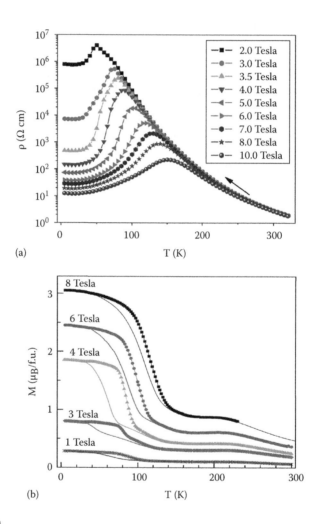

(a)

(b)

FIGURE 5.10
We show in (a) our resistivity data for Al-doped $Pr_{0.5}Ca_{0.5}MnO_3$ sample, which has a lower entropy FM state. (Taken from Kumar, D. et al., *J. Phys.: Condens. Matter*, 24, 386001, 2012.) No drop in the resistivity was seen in $H = 0$, and the drop becomes increasingly larger as H is increased from 2 Tesla. In (b), we show our magnetization data for the same Al-doped $Pr_{0.5}Ca_{0.5}MnO_3$ sample. (Taken from Banerjee, A. et al., *J. Phys.: Condens. Matter*, 21, 026002, 2009.) The rise seen in the magnetization becomes increasingly larger as H is increased. This data again confirms the prediction of Figure 5.5b.

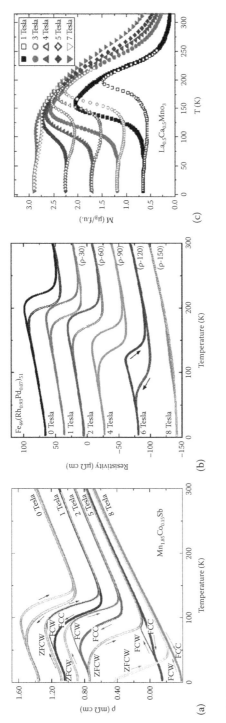

FIGURE 5.11

We show our resistivity and magnetization data for various samples, each of which has a lower entropy AFM state. A clear sharp hysteretic change in resistance and magnetization was seen in the lowest H, and the change becomes increasingly smaller as H is increased, becoming undetectable at the highest H used. This data confirms the prediction of Figure 5.5a. (a: Taken from Kushwaha, P. et al., *J. Phys. Condens Matter*, 20, 022204, 2008; b: Taken from Kushwaha, P. et al., *Phys. Rev. B*, 80, 174413, 2009; c: Taken from Dash, S. et al., *Phys. Rev. B*, 82, 172412, 2010.)

completion simply by changing the cooling field. We now discuss further manifestations of this new phenomenon.

5.3 Kinetic Arrest and Interrupted Transitions: Isothermal Variation of the Field

We now discuss one of the most frequent measurement protocols for magnetic materials, viz., isothermal variation of the magnetic field after cooling in $H = 0$. As in earlier discussions, we shall discuss the transformation and evolution of the two phases. Experimental confirmation shall come from measurement of physical properties that are distinct in the two phases. The properties usually measured in magnetic materials are magnetization and resistance.

We first consider the case of the lower entropy phase having higher magnetization. We cool the sample in $H = 0$ to below $T_k(0)$, which is higher than $T_U^*(H = 0)$ for the AF phase to be arrested. (Note that $T_k(0)$ is below $T_C(0)$ in all situations.) The sample is still in phase-2, and the kinetics is now arrested. On further cooling of the sample into, or below, the $T^*(H = 0)$ band, no transformation into the lower entropy FM phase occurs because we are below $T_k(0)$ and kinetics is already arrested. At this T_0, we isothermally increase H, as depicted in Figure 5.12f. At H_{UP} satisfying $T_k(H = H_{UP}) = T_0$, we cross the T_k line and de-arrest occurs. In the case of PCMAO, this could be referred to as a field-induced melting of the charge-ordered state, but it actually better corresponds to a field (or pressure) induced de-arrest (or devitrification) of an arrested (or glass-like) state. The material is now in its equilibrium phase, which is the lower entropy FM phase. No phase transition line is encountered on further increasing or decreasing H at this T_0, and the material remains in the FM phase on cycling H. We do cross the T_k line on the reducing-H cycle, but it has no observable effect as there is no underlying thermodynamic transformation. Figure 5.12a shows the observed M-H curve at 5 K on raising H from 0 to 14 Tesla and then lowering it to 0 and raising it back to 14 Tesla. The virgin curve on cooling in $H = 0$ and then raising H from 0 shows the initial response of the AFM phase; de-arrest of this starts after 6 Tesla. After reaching 14 Tesla, the cycling between 0 and 14 Tesla shows the response of the FM state. The devitrification from the arrested AFM phase to the equilibrium FM phase appears to occur over a range of H starting at about 6 Tesla, not at the unique H_0 described earlier. We shall come back to this observation of disorder-induced broadening of $T_k(H)$ into a band as we point out more evidence of devitrification (or de-arrest, the more commonly used term) being broad and not sharp.

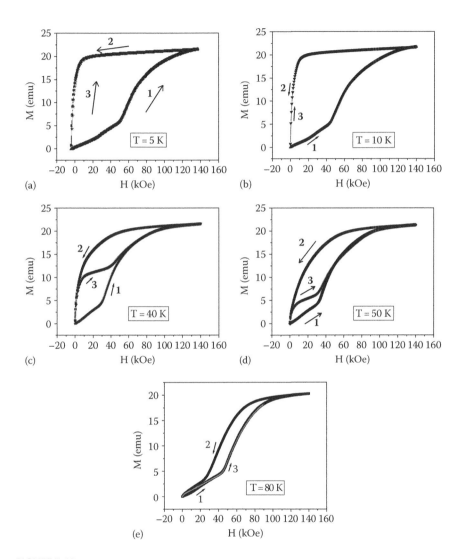

FIGURE 5.12
We show our isothermal M-H data for Al-doped PCMO sample at various temperatures. Parts of the data were reported by Banerjee and coworkers [18,21,25]. As T rises from (a) to (c), we note that the value of H at which there is a sharp rise in the initial curve (labeled 1) is reducing, but it is increasing at higher T in (d) and (e). Also, curve 3 is coming closer to curve 1, and they actually overlap for T = 70 K and above.   (*Continued*)

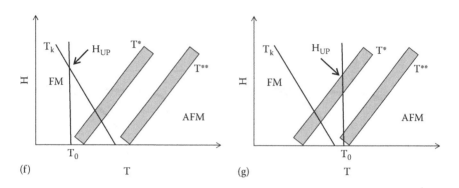

FIGURE 5.12 (*Continued*)
We show our isothermal M-H data for Al-doped PCMO sample at various temperatures. The initial curve 1 jutting out of the envelope curve 3 is an anomaly that was first observed and highlighted by Manekar et al. for Al-doped $CeFe_2$ [11] (except that the sample they studied has a lower entropy phase that was AFM, whereas this sample has a lower entropy phase that is FM). This is depicted in (f) by the decrease in H_{UP} as T_0 rises, and it also explains a fully FM phase in curves 2 and 3. The rise in M in response to increasing H occurs due to de-arrest of the arrested AFM phase; its occurrence over a range of H signifies that T_K is also a band like T* and not a sharp line. This supports the heuristic conjecture that if there is a distribution of T* and T**, there must also be a distribution of T_K. The behavior in (c) and (d) shows a partial conversion to the AFM phase on lowering H, and T_0 now lies in the region depicted in (g).

At a temperature that is slightly higher but still below T*(0), the T_K line will be crossed at a slightly lower H_{UP}, with no qualitative change in features. Figure 5.12b corresponds to a slightly higher T = 10 K, and we note that de-arrest now starts at below 5 Tesla. The virgin curve still corresponds to an AFM state, and the hysteresis curve is that of a fully FM state. The measured curves at T = 40 K and T = 50 K, shown in (c) and (d), depict that there is a partial back transformation to the AFM state on reducing H to 0, with the transformed fraction rising as T rises. This back transformation is complete at T = 80 K as shown in (e). Data in (c) and (d) correspond to the isothermal scans being at a temperature lying in the T**(H = 0) band, as shown in (g).

We now consider the case of the lower entropy phase having lower magnetization. We cool the sample in H = 0 to below the T* band and are still above $T_K(0)$. Phase-1 has completely transformed to the AFM phase-2. We cool further and cross $T_K(0)$. Since T_K rises with increasing H, we are at a temperature T_0 where isothermal variation of H will never cross T_K. We now start raising H isothermally at this T_0, which is lower than $T_K(0)$. As we increase H, we enter and cross the supercooling (T*, H*) band, and the AFM phase is still the equilibrium phase. The AFM phase will start transforming to the FM phase as we enter the (T**, H**) band, and the transition is complete as we cross this band. In Figure 5.13a, we show data at $T_0 = 5$ K for $La_{0.5}Ca_{0.5}MnO_3$ (LCMO) where this happens between 11 and 14 Tesla. The study on LCMO showed a fully FM phase at 14 Tesla. When H is now cycled between ±14 Tesla, the sample should transform to the AFM phase as H enters and crosses the T* band. T_0 is, however, lower than $T_k(H = 0)$ as we enter the T* band, and this reverse transformation is arrested. The material remains in the FM phase as H is cycled at T_0, and this was observed by Banerjee et al. [21] in the M-H curves of LCMO at 5 K. Banerjee et al. [26] also cooled to 5 K in various fields, viz., 2, 2.5, 3.5, 4.5, and 8 Tesla, and obtained various arrested fractions. These fractions did not change as the fields were reduced isothermally to the corresponding negative values, confirming that

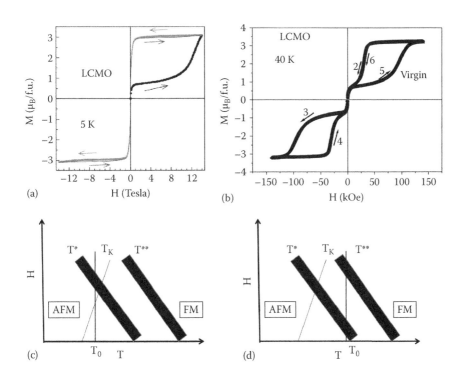

FIGURE 5.13

We show data from our isothermal M-H measurements for $La_{0.5}Ca_{0.5}MnO_3$ published in references [21,24,26,27]. Data in curve (a) is at T = 5 K and corresponds to a T_0 below $T_k(0)$. The envelope curve corresponds to the arrested FM state. The initial curve jutting out of the envelope curve is an anomaly that was first observed and highlighted by Manekar et al. for Al-doped $CeFe_2$ [11]. Data in (b) is at T = 40 K and corresponds to a T_0 as in schematic in (c) and (d).

(Continued)

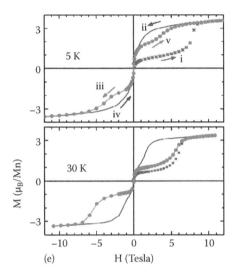

(e)   H (Tesla)

FIGURE 5.13 (*Continued*)

In (e), we show our data on $Nd_{0.5}Sr_{0.5}MnO_3$ (NSMO). The decreasing-H cycle (ii) clearly shows de-arrest below about 2 Tesla. More importantly, this de-arrest occurs over a range of H, consistent with Figure 5.12a and b. The question we raised was "would not the disordered system have a (H_k, T_k) band formed out of the quasi-continuum of (H_k, T_k) lines where each line would correspond to a local region of the sample?", and we now have convincing data that T_k is also a band like T* and not a sharp line. (a: Data from Banerjee, A. et al., *J. Phys. Condens Matter*, 21, 026002, 2009; e: Data from Rawat, R. et al., *J. Phys. Condens Matter*, 19, 256211, 2007.)

the T_k line was not being crossed at 5 K. The isothermal scan in (b) is at T = 40 K and shows a complete back transformation to the AFM phase as H is lowered to 0. This corresponds to the temperature being below the T* band at H = 0, as depicted in (c) and (d).

In Figure 5.13e, we show our data on $Nd_{0.5}Sr_{0.5}MnO_3$ (NSMO) from reference [28], at T_0 = 5 and 30 K. The decreasing-H cycle (ii) clearly shows a de-arrest below about 2 Tesla, and this de-arrest occurs over a range of H. We now have similar convincing data that T_K is also a band like T* and not a sharp line. We now replace the T_K line by a band, and show in Figure 5.14a and b new schematics that replace the corresponding schematics in Figure 5.4. This entire conceptual picture, of broadened T*, T_C, and T** bands and their interplay with the broadened kinetic arrest T_k band, was introduced with schematic figures by us in Manekar et al. [11]. All concepts were introduced simultaneously in 2001, whereas we have been introducing these one at a time here with some benefit of hindsight.

We can follow through the discussion on isothermal variation of H, as in Section 4.3.4, and the nonmonotonic dependence of the transition field will be seen in H_{dn} for the AFM lower entropy state. Our data on NSMO and on doped FeRh, taken from references [28,23], respectively, is shown in (c) and (d). This nonmonotonic behavior of the decreasing-H phase boundary had been

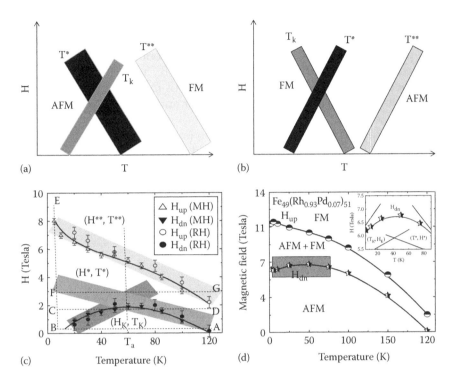

FIGURE 5.14
We consider T_k to be a broadened band in (a) and (b), instead of a line as considered in Figure 5.4a and b. We can follow through the discussion on isothermal variation of H, as in Section 4.3.3, and the nonmonotonic dependence of the transition field will be seen in H_{dn} for the lower entropy AFM state. (c: Taken from Rawat, R. et al., *J. Phys. Condens Matter*, 19, 256211, 2007; d: Taken from Kushwaha, P. et al., *Phys. Rev. B*, 80, 174413, 2009.)

observed in NSMO by Kuwahara et al. [15] in resistance measurements but was not understood as a consequence of kinetic arrest. This is important because this work [15] preceded Fath et al. [16] by 4 years, and the "interrupted phase transition" explanation could have precluded the proposition of an inhomogeneous ground state. In recent years also, similar nonmonotonic behavior has been reported in both magnetization and resistivity studies in magnetic shape memory alloys [6–9] by authors who are apparently aware of the kinetic arrest phenomenon, but it has not been explained as an interplay of supercooling and kinetic arrest bands.

The existence of such a nonmonotonic behavior of the increasing-H phase boundary, for a lower entropy FM state, was first observed by Wu et al. [19] in a single crystal sample of (La,Pr,Ca)MnO$_3$. They used a combination of isothermal measurements of resistivity and magnetization and explained it as an interplay between kinetic arrest and metastable states. The phase boundaries were attributed to de-arrest of the glass phase at low T and to the metastability limit at higher T. A similar nonmonotonicity of the increasing-H phase boundary has been reported in Ta-doped $HfFe_2$ by Rawat et al. [29], who attributed it to kinetic arrest of the higher entropy AFM phase on cooling in H = 0. A similar nonmonotonicity has also been reported recently by Fujieda et al. [30] in $LaFe_{12}B_6$, and by Bag et al. [31] in Nd-doped LaFe11.5Al1.5. The behavior in both these materials has again been attributed to kinetic arrest of the higher entropy AFM phase on cooling in H = 0.

Measurement of isothermal M-H curves (and of R-H curves) remains a very routine and popular measurement, and thus the nonmonotonic behaviors in the appropriate transition fields are important signatures of kinetic arrest or "interrupted phase transitions."

5.4 CHUF Protocol

We have discussed in Section 5.2 how by cooling in different values of H we can obtain states in which the higher entropy phase and the lower entropy phase coexist in different fractions at temperatures well below T*(H). The phase transition has been interrupted by the (T_k, H_k) line, and the higher entropy phase exists as an arrested state. This explains the puzzling phenomenon of phase coexistence to the lowest temperature and predicts continuous tunability of phase fractions, which has been subsequently observed as predicted. Further predictions were made on the T-dependence of the commonly measured isothermal M-H curves, and observations verifying these have been discussed in the previous section. We now discuss the behavior of these continuously tunable phase fractions on heating in different fields, under the recently conceived CHUF protocol discussed in Chapter 4.

In the discussion in Chapter 4, we had a (T*, H*) line, and this has now broadened into a band. In Figure 5.4, we indicated the characteristic H_L and H_U, or T_L and T_U, which were merged into one in the discussion in Chapter 4. We now have a coexistence of continuously variable fractions of phase-2 and phase-1 at any (H,T) point in the shaded regions shown in Figures 5.6 and 5.8, whereas we could have only either equilibrium phase-1 (100%) or arrested phase-2 (100%) at any point in this region in the absence of disorder broadening. In the discussion in Chapter 4, on heating in an appropriate H_W, the behavior depends on the inequality between $T_k(H_W)$ and T*(H_W). If T* is higher, then we would have arrested phase-2 transforming to equilibrium phase-1 at T_k and then to equilibrium phase-2 at T**, and thus exhibiting a reentrant transition. If T_k is higher than T*, then the phase-2 is de-arrested at T_k and sits in a local minimum as a metastable phase-2, and there is no transformation observed either at T_k or at T**. How has this scenario changed?

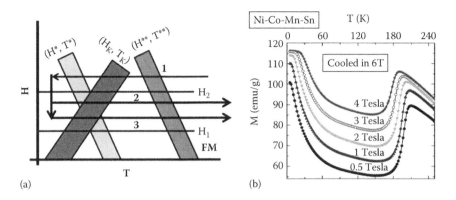

FIGURE 5.15
We follow on the discussion of the CHUF protocol in Section 4.3.2, and show the counterpart of Figure 4.8, with each of T*, T**, and T_K now being broad bands. The schematic in (a) follows our reference [18]. For cooling path 1, de-arrest and a reentrant transition will be seen for both paths 2 and 3; the extent of de-arrest will be more for path 3. Further, de-arrest will be seen at lower T for the lower warming field of path 3. We show our data on NiCoMnSn sample in figure (b), which also confirms these two results. (b: Taken from Lakhani, A. et al., *J. Phys. Condens Matter*, 24, 386004, 2012.)

The disorder broadening interrupts the broad phase transition on cooling in a field between H_L and H_U and completes the same broad reverse transition on crossing the T** band in the heating cycle, as shown in the data in Figures 5.10 and 5.11. The fraction that completed the transition on cooling is the one that will complete the reverse transition; the fraction arrested in phase-2 remains in phase-2 on getting de-arrested as it is above T* and is metastable, and again becomes the equilibrium phase on crossing T**. What happens if, as discussed earlier for the case of sharp first order transitions, we cause an isothermal change in H at the lowest temperature before warming? This is depicted in Figures 5.15 and 5.16, where we show cases of the lower entropy phase being AFM and FM, respectively. We presently assume that the T_k line is not crossed in this isothermal variation of H.

Let us consider the situation where we have cooled the sample in a field H_{Cool} between H_L and H_U so that we have phase coexistence, and we cool to a temperature T_0 well below the (H_k, T_k) line. We now vary H isothermally to H_{Heat} at T_0, but by an amount such that the (H_k, T_k) line is not encountered. This variation in H can be larger if T_0 is lower. There will be no change in the phase fraction during this process. With the field fixed at H_{Heat}, we now start heating the sample. Cooling in H_{Cool} and warming in H_{Heat} are what we referred to as the CHUF protocol. What do we expect?

On cooling in H_{Cool}, the transformed phase-1 fraction is $(T_U - T_k(H_{Cool}))/(T_U - T_L)$ and the arrested phase-2 fraction is $X(H_{Cool}) = (T_k(H_{Cool}) - T_L)/(T_U - T_L)$. Each of these temperatures T_U and T_L is actually a function of H but is not explicitly shown for simplicity. We consider in Figure 5.15 that the lower entropy phase has lower magnetization and assume first that H_{Heat}

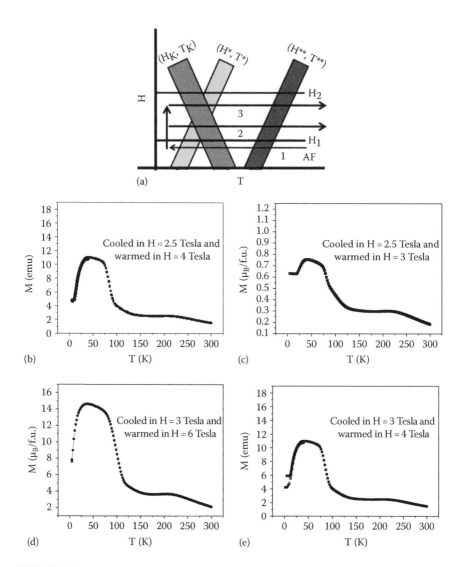

FIGURE 5.16
We follow on the discussion of the CHUF protocol in Section 4.3.2, and show the counterpart of Figure 4.8, with each of T*, T**, and T_K now being broad bands. The schematic for the lower entropy FM phase follows from our reference [18]. For cooling path 1, de-arrest and a reentrant transition will be seen for both paths 2 and 3; the extent of de-arrest will be more for path 3, and it will start at a lower temperature. Further, de-arrest will be seen at lower T as the warming field is lowered. We show our data from Banerjee et al. [21,25], for Al-doped $Pr_{0.5}Ca_{0.5}MnO_3$ in (b) and (c) for a cooling H = 2.5 Tesla and different warming fields of 3 and 4 Tesla, which confirm these two results. In (d) and (e), we show our results from Banerjee et al. [21,25], for a cooling H = 3 Tesla and different warming fields of 4 and 6 Tesla, which again confirm these two results.

is smaller than H_{Cool}. Here, T_U and T_L increase as H reduces while T_k falls as H reduces. So when we start heating in $H_{Heat} < H_{Cool}$, there is a larger region of the T* band lying above T_k, $(T_k - T_L)$ decreases, and X(H) will fall as H reduces until we reach H_L where $T_k = T_L$. As we raise temperature in field H_{Heat}, we enter the T* band from below, and there is no change in phase fractions as we are below $T_k(H_{Heat})$, but as we cross $T_k(H_{Heat})$, only a fraction (we continue here with T_k being a line as in Figure 5.4, and with the simplifying linear approximations) $X(H_{Heat}) = (T_k(H_{Heat}) - T_L(H_{Heat}))/(T_U(H_{Heat}) - T_L(H_{Heat}))$, which is below $T_k(H_{Heat})$, can stay arrested. We note that the fraction $X(H_{Cool}) - X(H_{Heat})$ gets de-arrested at $T = T_k(H_{Heat})$, and we observe a transformation from arrested FM to equilibrium AFM. The fraction $X(H_{Heat})$ will remain in the arrested FM state, and $(1 - X(H_{Heat}))$ is in the equilibrium AFM state. On further heating, this equilibrium AFM state will convert to the equilibrium FM phase as the T** band is crossed. So we observe a reentrant transition with a fraction $X(H_{Cool}) - X(H_{Heat})$ converting from FM to AFM at $T = T_k(H_{Heat})$ and the fraction $(1 - X(H_{Heat}))$ converting from AFM to FM across the $T^{**}(H_{Heat})$ band. The fractions undergoing the two transitions can be continuously controlled by a suitable choice of H_{Cool} and H_{Heat}. In the limiting case of $H_{Cool} > H_U$ and $H_{Heat} < H_L$, both steps correspond to full conversions, while more interesting cases emerge when both H_{Cool} and H_{Heat} lie between H_L and H_U. However, no de-arrest transition will be observed on crossing T_k if $H_{Heat} > H_{Cool}$. This is because $T_k(H_{Heat}) - T_L(H_{Heat})$ is then larger than $T_k(H_{Cool}) - T_L(H_{Cool})$ and no de-arrest will take place on crossing $T_k(H_{Heat})$. In Figure 5.15, we show the CHUF curves for different values of H_{Heat}.

Experimentalists worry about possible artifacts in measurement and like to keep the largest possible number of parameters unchanged if they need to compare two measurements. For this reason, visual contrasts under similar measurement conditions correspond to having a fixed value of H_{Heat} and varying H_{Cool} in a controlled way. In the first reports of CHUF measurements, made for first order magnetic transitions with the lower entropy phase having lower magnetization and higher resistance, Dash et al. [33] measured resistance in NSMO under the CHUF protocol with $H_{Heat} = 0$ and $H_{Cool} = 1$ and 3 Tesla. In both cases, H_1 was larger than H_{Heat}, and a reentrant transition was observed. They reported magnetization measurements for $H_{Heat} = 2$ Tesla and $H_{Cool} = 3$ Tesla when a reentrant transition was observed, and for $H_{Heat} = 2$ Tesla and $H_{Cool} = 0$ when only one transition, at $T^{**}(H = 2$ Tesla$)$, was observed. They also reported magnetization measurements for $H_{Heat} = 1$ Tesla with $H_{Cool} = 0$, 1, 2, and 3 Tesla. Only for $H_{Cool} = 2$ and 3 Tesla a reentrant transition was observed, and for $H_{Cool} = 0$ and 1 Tesla only one transition was observed at $T^{**}(H = 1$ Tesla$)$. The reentrant transition was observed while warming only if the measurement field is smaller relative to the cooling field. These were the first reported measurements of CHUF that comprehensively, and through the measurement of two different physical properties, confirmed that the kinetics of a broad first order magnetic transition was being arrested and the transition was "interrupted." This was soon followed by Banerjee et al. [21], reporting similar studies for the transition in PCMO, in which the lower entropy phase has higher magnetization. We shall discuss such transitions subsequently in this section. Here we continue with studies where the CHUF protocol was used, together with measurement of physical properties, to show kinetic arrest in various materials that showed a broad first order magnetic transition to a lower entropy phase that was AFM, or had lower magnetization than the higher entropy phase. Roy and Chattopadhyay [34] reported magnetization in Ru-doped $CeFe_2$ with the measurements in a warming field H_{Heat} = 2 Tesla and cooling fields H_{Cool} varying from 0 to 2.8 Tesla. They observed a reentrant transition for H_{Cool} = 2.8 and 2.5 Tesla but a single magnetization jump for H_{Cool} = 0, 1.7, 1.9, and 2 Tesla. Haldar et al. [35] later studied Si-doped $CeFe_2$ and reported magnetization with the measurements in a warming field H_{Heat} = 2 Tesla and cooling fields H_{Cool} varying from 0 to 3 Tesla. They observed a reentrant transition for H_{Cool} = 3 and 2.5 Tesla but a single magnetization jump for H_{Cool} = 0, 1.0, 1.5, and 2 Tesla. Since the whole concept of kinetic arrest of partially completed broad first order transitions had been proposed to explain the data for Al-doped $CeFe_2$, this should actually be viewed as a support for the phenomenology that has led to the idea of CHUF as a testing tool.

Kushwaha et al. [23] measured resistivity under the CHUF protocol in Pd-doped FeRh, which showed a broad but hysteretic FM to AFM transition on cooling. The T-dependence of the M-H curves showed nonmonotonic behavior in the transition field on the reducing-H cycle, as a confirmation of kinetic arrest. They measured resistance in a warming field H_{Heat} = 6 Tesla and cooling fields H_{Cool} varying from 0 to 8 Tesla. They observed a reentrant transition only for H_{Cool} = 8 Tesla, and this was confirmed by magnetization measurements. For H_{Cool} = 0 and 6 Tesla, only one transition of lowering of resistance was observed.

There was an initial report by Banerjee et al. [18] that can be considered as a precursor reflecting the development of the ideas behind the CHUF protocol, in which they measured magnetization in PSMO for a warming field H_{Heat} = 2 Tesla and reported a reentrant transition with a higher cooling field H_{Cool} = 3 Tesla. A more exhaustive study on LCMO, which also has an AFM phase with lower entropy, was reported by the same group just prior to their publication of the CHUF protocol in Dash et al. [33]. Chaddah et al. [27] reported magnetization showing a reentrant transition with a warming field H_{Heat} = 1 Tesla and a higher cooling field H_{Cool} = 6 Tesla, and resistance showing a reentrant transition with a warming field H_{Heat} = 0 and a cooling field H_{Cool} = 6 Tesla. Banerjee et al. [26] pursued this further with magnetization measurements in warming fields H_{Heat} = 3 and 1 Tesla and cooling fields H_{Cool} varying from 6 Tesla to 0. The data for H_{Heat} = 3 Tesla showed a reentrant transition for cooling fields H_{Cool} = 4.5 and 6 Tesla, and only one sharp rise in the magnetization for cooling fields H_{Cool} = 0, 1.5, and 3 Tesla. Sarkar et al. [36] reported studies on calcium-doped $YBaCo_2O_{5.5}$ cobaltites and concluded that "the cooling field can tune the fractions of the coexisting phases, and the glasslike state formed at low temperature can also be devitrified by warming the sample." They reported magnetization measurements under the CHUF protocol with a warming field H_{Heat} = 1 Tesla, and cooling fields H_{Cool} varying from 0 to 5 Tesla. They observed a reentrant transition for H_{Cool} = 3 and 5 Tesla but a single magnetization jump for H_{Cool} = 0, 0.2, 0.6, and 1 Tesla. They concluded that "The CHUF experimental protocol, clearly showing the devitrification of the arrested state, thus gives unambiguous and rather visual evidence of the coexisting phases in the magnetic glass state" [36].

Nayak et al. [37] had studied the Heusler compound Mn_2PtGa and found virgin curves outside the envelope curve in isothermal M-H measurements, with the anomaly becoming more prominent as temperature was lowered below 30 K. A first order FM to AFM transition was reported in this paper, but kinetic arrest was not considered as a possible explanation. Some months later, they communicated a study of magnetization with a warming field H_{Heat} = 1 Tesla and cooling fields H_{Cool} varying in small steps from 0 to 2 Tesla [38]. They observed a reentrant transition for H_{Cool} = 1.2, 1.5, and 2 Tesla but a single magnetization jump for H_{Cool} = 0.01, 0.1, 0.3, 0.5, 0.7, and 1 Tesla. They rightly concluded, following the predictions of the CHUF protocol [21], that the "observation of unequal magnetic behaviors when the sample is cooled with a field smaller/larger than the measuring field clearly indicates the kinetic arrest of the FM (FI) phase" [38].

Lakhani et al. [32] reported studies on a $Ni_{45}Co_5Mn_{38}Sn_{12}$ magnetic shape memory alloy. This class of materials shows a martensitic transition with a drop in magnetization on lowering T, and both resistivity and magnetization measurements indicated that the broad first order transition is kinetically arrested on cooling in higher H. Measurements on this sample by Banerjee et al. [39] and by Lakhani et al. [32] showed that arrest starts at a cooling field H_{Cool} = 3 Tesla and is complete for H_{Cool} = 9 Tesla. Measurements of magnetization were reported under the CHUF protocol with a warming field H_{Heat} = 4 Tesla and cooling fields H_{Cool} varying in steps of 1 Tesla from 0 to 8 Tesla, and with a warming field H_{Heat} = 1 Tesla and cooling fields H_{Cool} varying in steps of 1 Tesla from 0 to 5 Tesla. In this large set of reported data, a reentrant transition was seen only when $H_{Cool} > H_{Heat}$ and a single jump in magnetization was seen when $H_{Cool} \leq H_{Heat}$. To confirm the exhaustive initial measurements of Dash et al. [33] on NSMO, Lakhani et al. also measured magnetization with cooling field H_{Cool} fixed at 6 Tesla and warming field H_{Heat} varied from 0.5 to 4 Tesla, as also with cooling field H_{Cool} fixed at 3 Tesla and warming field H_{Heat} varied from 0.5 to 2 Tesla. In each of these cases, we had $H_{Cool} > H_{Heat}$, and reentrant transitions were observed. The qualitatively contrasting behavior for different signs of inequality between the cooling and warming fields was clearly established. The reentrant data for a larger cooling field, presented in some detail by both Lakhani et al. [32] and Nayak et al. [38], clearly shows that the de-arrest occurs over a broad range of temperatures.

As discussed earlier, the CHUF protocol with $H_{Cool} > H_{Heat}$ should give a reentrant phase transition, and this can be observed through the measurement of any physical property that varies sharply between the two phases. The most drastic of such observations was reported by Siruguri et al. [40] through neutron diffraction measurements. They measured neutron diffraction from an LCMO sample by cooling it from 300 to 9 K in $H_{Cool} = 7$ Tesla and then measure the diffraction patterns as the sample is warmed in $H_{Heat} = 0.5$ Tesla. They repeated the process with $H_{Heat} = 3$ Tesla and observed a reentrant transition in both cases. They also studied a $Ni_{37}Co_{11}Mn_{42.5}Sn_{9.5}$ sample, which was, again, cooled each time from 300 to 9 K in $H_{Cool} = 7$ Tesla and warmed in various $H_{Heat} = 0.5$, 1.0, 1.5, and 2 Tesla. Each run showed a reentrant transition, confirming kinetic arrest in both samples.

We now consider, in Figure 5.16, the case when the lower entropy phase is FM, or has higher magnetization, and H_{Heat} is larger than H_{Cool}. Here, T_U and T_L increase while T_k falls as H increases. (We continue here with T_k being a line, as in Figure 5.4, and with the simplifying linear approximations.) So when we start heating in $H_{Heat} > H_{Cool}$, there is a larger region of the T* band lying above T_k, $(T_k - T_L)$ decreases, and X(H) will fall as H increases until we reach H_U where $T_k = T_L$. As we raise temperature in field H_{Heat}, we enter the T* band from below, and there is no change in the phase fractions as we are below $T_k(H_{Heat})$, but as we cross $T_k(H_{Heat})$, only a fraction $X(H_{Heat}) = (T_k(H_{Heat}) - T_L(H_{Heat}))/(T_U(H_{Heat}) - T_L(H_{Heat}))$ that is below $T_k(H_{Heat})$ can stay arrested. We note that the fraction $X(H_{Cool}) - X(H_{Heat})$ gets de-arrested at $T = T_k(H_{Heat})$, and we observe a transformation from the arrested AFM to an equilibrium FM. The fraction $X(H_{Heat})$ will remain in the arrested AFM state, and $(1 - X(H_{Heat}))$ is in the equilibrium FM state. On further heating, this equilibrium FM state will convert to the equilibrium AFM phase as the T** band is crossed, so we observe a reentrant transition, with the fraction $X(H_{Cool}) - X(H_{Heat})$ converting from AFM to FM at $T = T_k(H_{Heat})$ and the fraction $(1 - X(H_{Heat}))$ converting from FM to AFM across the $T^{**}(H_{Heat})$ band. The fractions undergoing the two transitions can be continuously controlled by suitable choice of H_{Cool} and H_{Heat}. In the limiting case of $H_{Heat} > H_U$ and $H_{Cool} < H_L$, both steps correspond to full conversions, while more interesting cases emerge when both H_{Cool} and H_{Heat} lie between H_L and H_U. However, no de-arrest transition will be observed on crossing T_k if $H_{Heat} < H_{Cool}$. This is because $T_k(H_{Heat}) - T_L(H_{Heat})$ is then larger than $T_k(H_{Cool}) - T_L(H_{Cool})$ and no de-arrest will take place on crossing $T_k(H_{Heat})$.

The case of a lower entropy FM phase has been experimentally studied in great detail by Banerjee et al. [18,21,25] in PCMAO, with some subsequent studies of the CHUF in LPCMO [41] and in Ta-doped $HfFe_2$ [29]. We start by discussing the results on PCMAO. The initial report by Banerjee et al. [18], which was labeled as a precursor reflecting the development of the ideas behind the CHUF protocol, reported magnetization in PCMAO for a warming field H_{Heat} = 4 Tesla, a reentrant transition with a lower cooling field H_{Cool} = 3 Tesla, and a single FM to AFM transition with a higher cooling field H_{Cool} = 6 Tesla. A more exhaustive study was reported by the same group [25], where they reported both magnetization and resistivity studies with the warming field H_{Heat} = 4 Tesla. In magnetization measurements, they used cooling fields H_{Cool} = 0, 2.5, 2.75, 3, 3.25, 3.5, 3.75, and 4 Tesla, all of which showed a reentrant transition. They also used higher cooling fields H_{Cool} = 5, 6, and 8 Tesla, all of which showed a single FM to AFM transition. In resistance measurements, they used cooling fields H_{Cool} = 0, 3, 3.5, and 4 Tesla, all of which showed a reentrant transition. They also used higher cooling fields H_{Cool} = 5, 6, and 8 Tesla, all of which showed a single FM to AFM transition. While the measurements comprehensively covered what was later christened the CHUF protocol, they were not appropriately labeled. The subsequent paper [21] carefully established the difference observed between the cases $H_{Cool} > H_{Heat}$, and $H_{Cool} \leq H_{Heat}$. They reported results for two sets of cooling field H_{Cool} (2.5 and 3 Tesla) with different values of warming field H_{Heat}, and for two sets of warming field H_{Heat} (3 and 4 Tesla) with different values of cooling field H_{Cool}. The predicted behavior under the CHUF protocol was now established for the higher entropy arrested phase having both higher and lower magnetization compared to the lower entropy equilibrium phase.

Following the identification by Wu et al. [19] of the kinetic arrest of the higher entropy insulating AFM phase and its coexistence with the lower entropy metallic FM phase in LPCMO, Sathe et al. [41] studied thin films of this material on different substrates, as providing different strain due to varying lattice mismatch. They reported measurement of resistance under CHUF protocol in a warming field H_{Heat} = 2 Tesla, with the cooling fields H_{Cool} = 1 and 4 Tesla. They observed a reentrant transition for H_{Cool} = 1 Tesla, and a single transition with resistance increase for H_{Cool} = 4 Tesla. A reentrant transition was also shown for one substrate with the cooling field H_{Cool} = 0.5 Tesla and warming fields H_{Heat} = 1, 1.5, and 2 Tesla.

More recently, Rawat et al. [29] reported studies on Ta-doped $HfFe_2$ where an AFM to FM transition occurs on lowering temperature and the zero-field transition temperature reduces with Ta-doping. They reported kinetic arrest for $Hf_{1-x}Ta_xFe_2$ (x = 0.225, 0.230, and 0.235) samples and its confirmation through magnetization measurements following the CHUF protocol. Reddy et al. [42] also collected Mossbauer spectra in the x = 0.230 sample, using the CHUF protocol, and observed a reentrant transition.

Table 5.1 lists all the materials in which kinetic arrest of a first order magnetic transition has already been reported.

TABLE 5.1

List of Materials Where a Kinetic Arrest Has Been Reported

1.	Doped $CeFe_2$; with the dopants (Al, Ru, Si, Ge)
2.	Gd_5Ge_4
3.	Doped Mn_2Sb; with the dopants (Co, Sn [43])
4.	A large number of 'half-doped' CMR manganites
5.	Nd_7Rh_3 [44]
6.	Doped FeRh; with the dopants (Pd, Ni [45])
7.	NiMnIn [46], NiMnSn, NiMnGa, based magnetic shape memory alloys
8.	Ca-doped YBaCoO
9.	Ta-doped $HfFe_2$
10.	Ba-doped $SmCrO_3$ [47]
11.	Mn_2PtGa
12.	$HoFe_4Ge_2$ [48]
13.	Se-doped CoS_2 [49]
14.	$FeAl_2O_4$ [50]
15.	$LaFe_{12}B_6$

This long, and continuously growing, list shows the ubiquity of "kinetic arrest" across first order magnetic transitions.

Most, but not all, of these materials have been discussed here. Additional essential references are noted in the case of materials that have not been discussed in this book.

5.5 Measuring $T_k(H)$ Using CHUF

We have seen earlier that the arrested fraction gets de-arrested under the CHUF protocol, provided the warming field is lower than the cooling field for the case of the lower entropy phase being AFM and provided the warming field is higher than the cooling field for the case of the lower entropy phase being FM. This de-arrest occurs at $T_k(H_{Heat})$, and by varying H_{Heat} we can measure this H-dependence. We first consider the case $H_{Cool} < H_L$ for the case of the FM ground state, and $H_{Cool} > H_U$ for the AFM ground state, so that the initial state is the fully arrested higher entropy phase. Then, we isothermally vary the field to H_0 lying between H_L and H_U and increase T in this field H_0. As we cross the line $T_k(H)$, the arrested higher entropy phase will get partially de-arrested to the lower entropy phase, with the fraction $(T_k(H_0) - T_L(H_0))/(T_U(H_0) - T_L(H_0))$ still remaining arrested in the higher entropy phase. (It should be mentioned that the highly useful glass ceramics are also formed by partial crystallization on heating of a glass.) This de-arrest at $T_k(H_0)$ will be visible as a sharp change in the relevant physical property as the material transforms from a fully higher entropy phase to a state with the coexistence of two phases and allows a measurement of the $T_k(H)$ line. Their structural counterpart would correspond to a measurement of $T_g(P)$ at various pressures—a quantity that remains tortuous to measure.

We have mentioned earlier that de-arrest during isothermal variation of H in an appropriate temperature regime is not observed as a sharp transition, and raised the query that if the phase transition is broadened by disorder, then the temperature for kinetic arrest may also not remain sharp. We shall present inconvertible evidence supporting this view now. We consider the cooling field H_{Cool} to correspond to a larger fraction of the arrested phase, and the warming field H_0 to correspond to a smaller fraction of the arrested phase, so that de-arrest occurs on heating. The de-arrest following this CHUF protocol should occur, on heating in a field H_0, through a sharp transformation at $T_k(H_0)$. Measurements under this CHUF protocol for various materials, with both FM and AFM ground states, however, showed that de-arrest is initiated with a sharp step at a temperature that also depends on H_{Cool}. Examples are the detailed reports on NiCoMnSn [32,39] as also on PCMAO [25], as shown in Figure 5.17. This clearly brings out that we do not have a sharp T_k at H_0. We rather have a range of T, at a given warming field H_0, over which de-arrest occurs. It follows that the arrest during cooling in a fixed field would also occur over a range of temperatures. This brings us back to the point raised in Section 5.2, viz., "If such a kinetic arrest were to occur below a (H_k, T_k) line in the pure system, would not the disordered system have a (H_k, T_k) band formed out of the quasi-continuum of (H_k, T_k) lines where each line would correspond to a local region of the sample?" We note from the magnetization data on PCMAO that for $H_0 = 4$ Tesla, and as H_{Cool} is varied from 2.5 to 3.75 Tesla, the temperature T_k, at which the sharp step of de-arrest starts, increases from about 15 K to about 25 K. Similarly, the magnetization data on NiCoMnSn for $H_0 = 4$ Tesla shows that as H_{Cool} is varied from 8 to 5 Tesla, T_k

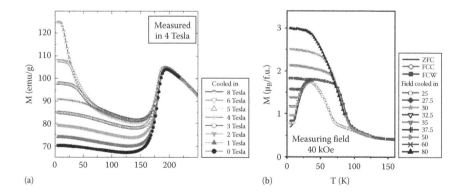

FIGURE 5.17
We show CHUF measurements in a fixed $H_{Heat} = 4$ Tesla. Figure (a) corresponds to our data for a lower entropy AFM state, while Figure (b) corresponds to our data for a lower entropy FM state. De-arrest is seen for H_{Cool} above 4 Tesla in (a) and for H_{Cool} below 4 Tesla in (b). The de-arrest temperature changes with H_{Cool}, bringing out that T_k is a band, because if $T_k(H)$ is a line, then the de-arrest temperature can only depend on H_{Heat} and not on H_{Cool}. (a: Taken from Lakhani, A. et al., *J. Phys. Condens Matter*, 24, 386004, 2012; b: Banerjee, A. et al., *Phys. Rev. B*, 74, 224445, 2006.)

increases from about 15 K to about 50 K. Further, the magnetization data on NiCoMnSn for $H_0 = 1$ Tesla shows that as H_{Cool} is varied from 5 to 2 Tesla, T_k increases from about 12 K to about 40 K. This clearly answers the earlier question and shows that different regions that have different values of T* also have different values of T_k. If we could figure out a correlation between the T* and the T_k of regions with different amounts of frozen (or quenched) disorder, it could help understand the physics of glass formation.

This has been considered by Chaddah et al. [12] by assuming two simple but extreme possibilities, namely, the region with the highest T* has the lowest T_k, i.e., the two bands are anticorrelated, and the region with the highest T* has the highest T_k, i.e., the two bands are correlated. They studied the phenomenology of de-arrest under isothermal variation of H and that under warming, with the appropriate CHUF protocol. The two extreme cases predicted behavior with differences that were visually drastic [12–14]. These predictions have been tested in many materials showing kinetic arrest and an interrupted first order magnetic transition [13,14,21,22,26–29,31,49,51,52]. In all but reference [52], anticorrelation has been observed between the T_k and T* bands. This anticorrelation has been naively understood in the following manner. A disorder lowers the transition temperature of the first order transition because it lowers long-range correlations but enhances the possibility of arresting the kinetics because it creates hindrances. In the last case, the studies were on nanoparticles where a disorder emanates from the lack of LRO on the surface and increases as size reduces or the fraction of atoms on the surface rises. This increase of open space due to larger fraction on the surface, however, also reduces kinetic hindrance and lowers T_k. This naive explanation needs to be supported by more rigorous theory.

5.6 What Causes Kinetic Arrest?

We now attempt to understand what factors help arrest a first order magnetic transition. We recognize that we are indirectly addressing the long-standing question "why do glasses form?" [53], in much the same way as we studied doped-$CeFe_2$ in an attempt to understand the puzzling data in RSMO crystals.

We first stress that some of these magnetic transitions are not accompanied by any structural change. The transition is not accompanied by diffusive motions, and the arrest of the transition cannot be convincingly related to diffusivity or viscosity. Is there another criterion that decides when a first order transition is arrested? Any new criterion we propose must not counter the diffusivity and viscosity criteria that have been so exhaustively tested in the arrest of structural transitions.

We have recognized in the initial chapters that the time for completion of any first order phase transition is governed by the time required to extract the latent heat of the system. This is another time scale for kinetics that we propose can be used as an alternative to diffusion time for completion of the transition.

In a usual cooling process in the lab, heat is conducted away by thermal conduction. In metallic systems at low temperature, this thermal conduction is dominated by conduction electrons. We are aware that at very low temperatures local magnetic moments are weakly coupled to the lattice, and the nuclear demagnetization technique for cooling does produce nuclear temperatures around 50 nK, which are much lower than the electron temperatures of around 50 μK in the same material. We are also aware that thermal equilibrium among nuclei is reached in a time characterized by the spin–spin relaxation time, whereas thermal equilibrium between nuclear spins and conduction electrons takes many hours at these temperatures as it is governed by the spin–lattice relaxation time. The thermal equilibration times are very large because of the long relaxation time, and two distinct temperatures can be identified simultaneously in the same specimen. We have argued that removal of specific heat and latent heat can similarly be identified by two different characteristic times.

The latent heat of first order magnetic transitions corresponds to ordered magnetic structures, and this magnetic order could be due to localized orbital magnetism or itinerant electron magnetism. It is obvious that the latter would have stronger coupling to conduction electrons than the former. Thermal conduction would remove latent heat corresponding to itinerant electron magnetism faster than it would remove latent heat corresponding to orbital magnetism. The magnetic transition caused by the latter is, thus, more likely to be kinetically arrested.

Have we violated any concept corresponding to the formation of a structural glass by kinetic arrest of the first order freezing transition? We reproduce the succinct comments in Lakhani et al. [32] in this context: "In structural glasses the liquid to crystal freezing is arrested, while in magnetic glasses the magnetic re-ordering is arrested. In both cases the material is cooled through the first order transition temperature without extracting latent heat. While in a jamming process (only) the translational kinetics is arrested, in the formation of both these glasses the specific heat is extracted, while the latent heat (which has a different coupling with the thermal conduction process) cannot be extracted [54]. In both cases the higher entropy phase persists down to the lowest temperature [55]. In this sense, the magnetic glass and the structural glass address similar physics. However, in structural glasses the pressure is the variable parameter. The experimental advantage of magnetic field (H) over pressure (P) is that it can be easily varied reversibly without a control medium, and its variation does not interfere with temperature control." (The reference numbers in the original text being quoted were different; they have been changed as per the present listing of references.)

The only test of this concept reported so far is by Rawat et al. [29]. They presented their motivation as follows: "Recently, Chaddah and Banerjee [54] have proposed that magnetic glasses form in these systems when magnetic latent heat is weakly coupled to a thermal conduction process. In $Hf_{1-x}Ta_xFe_2$, the first-order transformation is accompanied by significant latent heat [56] and the first-order transition depends sensitively on Ta content [57]. In contrast to magnetic coupling, the conduction electron mediated coupling is not expected to change much, as the lattice structure remains unchanged with small variations in lattice parameter. Therefore, the arrest of kinetics is expected to dominate with increasing x. Here, we present our results on Ta-doped $HfFe_2$, for which first-order transition temperature is tuned to near $T = 0$ K." Based on their measurements on x varying from 0.225 to 0.235, they concluded that "the monotonic rise in H_k as x varies from 0.225 to 0.235, shows that the formation of a magnetic glass becomes more favorable with increasing x. As noted earlier, the lattice parameters decrease monotonically as x increases. This would favor a direct exchange magnetic coupling, as the wavefunction overlap would rise exponentially, over a conduction electron mediated coupling (that would not vary much with interatomic spacing). Therefore, our results could be relevant to the proposal of Chaddah and Banerjee [54] that a magnetic glass is formed when the magnetic latent heat is weakly coupled (cf the sample specific heat) to the thermal conduction process." (Here also, the reference numbers in the original text being quoted were different; they have been changed as per the present listing of references.)

Clearly, the proposal to link the coupling of latent heat to thermal conductivity as the criterion for answering "why do glasses form?" has not yet been negated. It opens a new window to help understand why structural glasses form. This needs to undergo further tests.

References

1. Y. Imry and M. Wortis, *Phys Rev B* **19** (1979) 3580.
2. P.N. Timonin, *Phys Rev B* **69** (2004) 092102.
3. M.K. Chattopadhyay, S.B. Roy, A.K. Nigam, K.J.S. Sokhey, and P. Chaddah, *Phys Rev B* **68** (2003) 174404.
4. A. Soibel, E. Zeldov, M. Rappaport, Y. Myasoedov, T. Tamegai, S. Ooi, M. Konczykowski, and V.B. Geshkenbein, *Nature* **406** (2000) 283.
5. S.B. Roy, G.K. Perkins, M.K. Chattopadhyay, A.K. Nigam, K.J.S. Sokhey, P. Chaddah, A.D. Caplin, and L.F. Cohen, *Phys Rev Lett* **92** (2004) 147203.
6. W. Ito, K. Ito, R.Y. Umetsu, R. Kainuma, K. Koyama, K. Watanabe, A. Fujita, K. Oikawa, K. Ishida, and T. Kanomata, *Appl Phys Lett* **92** (2008) 021908.
7. R.Y. Umetsu, W. Ito, K. Ito, K. Koyama, A. Fujita, K. Oikawa, T. Kanomata, R. Kainuma, and K. Ishida, *Scr Mater* **60** (2009) 25.

8. R.Y. Umetsu, K. Endo, A. Kondo, K. Kindo, W. Ito, X. Xu, T. Kanomata, and R. Kainuma, *Mater Trans* **54** (2013) 291.

9. X. Xu, W. Ito, M. Tokunaga, T. Kihara, K. Oka, R.Y. Umetsu, T. Kanomata, and R. Kainuma, *Metals* **3** (2013) 298.

10. A. Lakhani, S. Dash, A. Banerjee, P. Chaddah, X. Chen, and R.V. Ramanujan, *Appl Phys Lett* **99** (2011) 242503.

11. M.A. Manekar, S. Chaudhary, M.K. Chattopadhyay, K.J. Singh, S.B. Roy, and P. Chaddah, *Phys Rev B* **64** (2001) 104416.

12. P. Chaddah, A. Banerjee, and S.B. Roy. Correlating supercooling limit and glass-like arrest of kinetics for disorder-broadened 1st order transitions: relevance to phase separation, http://arxiv.org/pdf/cond-mat/0601095 (2006). Accessed January 5, 2006.

13. K. Kumar, A.K. Pramanik, A. Banerjee, P. Chaddah, S.B. Roy, S. Park, C.L. Zhang, and S.-W. Cheong, *Phys Rev B* **73** (2006) 184435.

14. S.B. Roy, M.K. Chattopadhyay, A. Banerjee, P. Chaddah, J.D. Moore, G.K. Perkins, L.F. Cohen, K.A. Gschneidner, Jr., and V.K. Pecharsky, *Phys Rev B* **75** (2007) 184410.

15. H. Kuwahara, Y. Tomioka, A. Asamitsu, Y. Moritomo, and Y. Tokura, *Science* **270** (1995) 961.

16. M. Fäth, S. Freisem, A.A. Menovsky, Y. Tomioka, J. Aarts, and J.A. Mydosh, *Science* **285** (1999) 1540.

17. C. Magen, L. Morellon, P.A. Algarabel, C. Marquine, and M.R. Ibarra, *J Phys: Condens Matter* **15** (2003) 2389.

18. A. Banerjee, A.K. Pramanik, K. Kumar, and P. Chaddah, *J Phys: Condens Matter* **18** (2006) L605.

19. W. Wu, C. Israel, N. Hur, S. Park, S.-W. Cheong, and A. de Lozanne, *Nat Mater* **5** (2006) 881.

20. D. Kumar, K. Kumar, A. Banerjee, and P. Chaddah, *J Phys: Condens Matter* **24** (2012) 386001.

21. A. Banerjee, K. Kumar, and P. Chaddah, *J Phys: Condens Matter* **21** (2009) 026002.

22. P. Kushwaha, R. Rawat, and P. Chaddah, *J Phys: Condens Matter* **20** (2008) 022204.

23. P. Kushwaha, A. Lakhani, R. Rawat, and P. Chaddah, *Phys Rev B* **80** (2009) 174413.

24. S. Dash, K. Kumar, A. Banerjee, and P. Chaddah, *Phys Rev* **B82** (2010) 172412.

25. A. Banerjee, K. Mukherjee, K. Kumar, and P. Chaddah, *Phys Rev B* **74** (2006) 224445.

26. A. Banerjee, K. Kumar, and P. Chaddah, *J Phys: Condens Matter* **20** (2008) 255245.

27. P. Chaddah, K. Kumar, and A. Banerjee, *Phys Rev* **B77** (2008) 100402.

28. R. Rawat, K. Mukherjee, K. Kumar, A. Banerjee, and P. Chaddah, *J Phys: Condens Matter* **19** (2007) 256211.

29. R. Rawat, P. Chaddah, P. Bag, P.D. Babu, and V. Siruguri, *J Phys: Condens Matter* **25** (2013) 066011.

30. S. Fujieda, K. Fukamichi, and S. Suzuki, *J Mag Mag Mater* **421** (2017) 403.

31. P. Bag and R. Nath, *J Mag Mag Mater* **426** (2017) 525.

32. A. Lakhani, A. Banerjee, P. Chaddah, X. Chen, and R.V. Ramanujan, *J Phys: Condens Matter* **24** (2012) 386004.

33. S. Dash, A. Banerjee, and P. Chaddah, *Solid State Commun.* **148** (2009) 336.

34. S.B. Roy and M.K. Chattopadhyay, *Phys Rev B* **79** (2009) 052407.

35. A. Haldar, K.G. Suresh, and A.K. Nigam, *Intermetallics* **18** (2010) 1772.
36. T. Sarkar, V. Pralong, and B. Raveau, *Phys Rev B* **83** (2011) 214428 *Phys Rev B* **84** (2011) 059904(E).
37. A.K. Nayak, M. Nicklas, S. Chadov, C. Shekhar, Y. Skourski, J. Winterlik, and C. Felser, *Phys Rev Lett* **110** (2013) 127204.
38. A.K. Nayak, M. Nicklas, C. Shekhar, and C. Felser. Kinetic arrest related to a first-order ferrimagnetic to antiferromagnetic transition in the Heusler compound Mn_2PtGa, *arXiv* **1304**(4459) (2013) *J Appl Phys* **113** (2013) 17E308.
39. A. Banerjee, S. Dash, A. Lakhani, P. Chaddah, X. Chen, and R.V. Ramanujan, *Solid State Commun* **151** (2011) 971.
40. V. Siruguri, P.D. Babu, S.D. Kaushik, A. Biswas, S.K. Sarkar, M. Krishnan, and P. Chaddah, *J Phys: Condens Matter* **25** (2013) 496011.
41. V.G. Sathe, A. Ahlawat, R. Rawat, and P. Chaddah, *J Phys: Condens Matter* **22** (2010) 176002.
42. V.R. Reddy, R. Rawat, A. Gupta, P. Bag, V. Siruguri, and P. Chaddah, *J. Phys: Condens Matter* **25** (2013) 316005.
43. Y.Q. Zhang, Z.D. Zhang, and J. Aarts, *Phys Rev B* **70** (2004) 132407 *Phys Rev B* **71** (2005) 229902(E).
44. K. Sengupta and E.V. Sampathkumaran, *Phys Rev B* **73** (2006) 020406.
45. M.A. Manekar, M.K. Chattopadhyay, and S.B. Roy, *J Phys: Condens Matter* **23** (2011) 086001.
46. V.K. Sharma, M.K. Chattopadhyay, and S.B. Roy, *Phys Rev B* **76** (2007) 140401.
47. X.L. Qian, D.M. Deng, Y. Jin, B. Lu, S.X. Cao, and J.C. Zhang, *J Appl Phys* **115** (2014) 193902.
48. J. Liu, V.K. Pecharsky, and K.A. Gschneidner, Jr., *J Alloys Comp* **631** (2015) 26.
49. S.K. Mishra and R. Rawat, *State Commun* **244** (2016) 33.
50. H.S. Nair, R. Kumar, and A.M. Strydom, *Phys Rev B* **91** (2015) 054423.
51. K. Mukherjee, K. Kumar, A. Banerjee, and P. Chaddah, *Eur Phys J* **B86** (2013) 21.
52. R. Rawat, P. Chaddah, P. Bag, K. Das, and I. Das, *J Phys: Condens Matter* **24** (2012) 416001.
53. L. Berthier and M. Ediger, *Phys Today* **69** (2016) 40.
54. P. Chaddah and A. Banerjee. Magnetic Glass formed by kinetic arrest of first order phase transitions, *arXiv* **1004**(3116v3) (2012).
55. A. Banerjee, R. Rawat, K. Mukherjee, and P. Chaddah, *Phys Rev B* **79** (2009) 212403.
56. H. Wada, N. Shimamura, and M. Shiga, *Phys Rev B* **48** (1993) 10221.
57. Y. Nishihara and Y. Yamaguchi, *J Phys Soc Jpn* **52** (1983) 3630.

Index

A

Abrikosov lattice, 17, 64, 68
Absolute minimum, 40–46, 50–53, 62, 76
Amorphization, 95
Anti-correlation, 148
Arrest, of kinetics, of transition, 39, 40, 75, 81–83, 86–111, 119–150
Arrhenius law, 86
Artifact, 43, 44, 65, 67, 70, 71, 81–83, 114, 141
Astronomical time-scale, 90

B

Bands, 113, 117, 121, 123, 136–140, 148
Barrier, 40–52, 65, 77–80, 88–94, 100, 103
Bean's critical state model, 72
Boiling, , 2–5, 52, 81
Brain, 1, 2
Brillouin zone, 81
Broad transition, 36, 43, 64–69, 108, 111, 113–147

C

CeFe$_2$, 81, 117, 123, 124, 126, 142, 146, 148
   Al-doped, 81, 98–100, 108, 134, 135, 142, 146
   Ru-doped, 81, 92, 93, 100, 115, 116, 142, 146
Charge order, 53, 73, 75, 97, 132
Clausius-Clapeyron, 15–21, 35, 43, 52, 53, 63–69, 73, 80, 97
Co-doped Mn$_2$Sb, 75, 81, 97–100
Coercive field, 70
Coexisting phase fractions, 73, 117–129, 143–146
Coherence volume, 17, 52, 78
Colossal magnetoresistance (CMR) manganite, 73, 122, 123, 146
Compressibility, 7, 13–15, 23–26, 30, 48
Continuous phase transition, 36–43, 47, 50, 70, 72, 82

Continuously tunable, 128, 138
Control variables, 10–14, 19–23, 29–32, 43, 48, 50–62, 63–65, 68–72, 75–83, 90–95, 102, 106, 110, 111, 113, 117, 119, 120
Cooling and heating in unequal fields (CHUF), 101–106, 110, 126, 138–148
Cooper pairs, 17, 52
Correlation length, 36, 39, 52, 66–69, 113
Critical cooling rate, 87, 94, 98, 103
Critical current density, 64, 72, 77, 85
Critical point, 4, 18, 27–31
Critical State model, 71, 72
Crystallize, 40, 52, 85, 87–91, 146
Curie point, 35

D

Devitrification, 102, 132, 143
Diamagnetism, 7, 8, 17
Dielectric, 10, 54, 91, 97, 102
Diffusivity, 85–87, 148, 149
Domains, 70–72, 83

E

Ehrenfest classification, 14–18, 23, 32–46, 53, 63–70, 78, 115
Electrical resistance, 7, 9, 19, 97
Envelope hysteresis curve, 81, 82, 107–111, 134, 135, 143
Equilibrium phase, 9, 11, 12, 20, 31, 33, 39, 90, 96, 97, 102, 124, 132, 134, 138, 139, 145
Equilibrium physical property, 63, 66
Exchange bias, 54

F

FeRh, 22, 100, 107, 108, 136, 142, 146
Fermi liquid, 9
Fermi surface, 97